P9-AOI-035

Agates

Treasures of the Earth

Roger Pabian with Brian Jackson,
Peter Tandy & John Cromartie

RIVERHEAD FREE LIBRARY
330 COURT STREET
RIVERHEAD, NEW YORK 11901

FIREFLY BOOKS

A FIREFLY BOOK

Published by Firefly Books Ltd. 2006

Copyright © 2006 Natural History Museum, London

All rights reserved. No part of this publication may be reproduced, stored in a retrieval system, or transmitted in any form or by any means, electronic, mechanical, photocopying, recording or otherwise, without the prior written permission of the Publisher.

First printing

Publisher Cataloging-in-Publication Data (U.S.)

Pabian, Roger K.
  Agates : treasures of the earth / Roger Pabian ; with Brian Jackson, Peter Tandy, and John Cromartie.
[192] p. : col. photos. ; cm.
Includes bibliographical references and index.
Summary: An illustrated resource to agates including an introduction to their geology and formation, a comprehensive identification guide and listings for where they can be found around the world.
ISBN-13: 978-1-55407-098-5
ISBN-10: 1-55407-098-8
1. Agates. I. Jackson, Brian. II. Tandy, Peter. III. Cromartie, John. IV. Title.
553.8/7 dc22   QE391.Q2.P335 2006

Library and Archives Canada Cataloguing in Publication

Pabian, Roger K
  Agates : treasures of the earth / Roger Pabian ; contributors, Brian Jackson, Peter Tandy and John Cromartie.
Includes bibliographical references and index.
ISBN-13: 978-1-55407-098-5
ISBN-10: 1-55407-098-8
  1. Agates. I. Title.
QE391.Q2P32 2006   553.   C2006-900613-X

Published in the United States by
Firefly Books (U.S.) Inc.
P.O. Box 1338, Ellicott Station
Buffalo, New York 14205

Published in Canada by
Firefly Books Ltd.
66 Leek Crescent
Richmond Hill, Ontario L4B 1H1

Front cover: *top* agate from Germany; *bottom* agate from Scotland.
Back cover: *top* faulted agate; *bottom* agate from Germany.
Title page: agate from Russian arctic coast.

Designed by Mercer Design

Printed in Singapore

Contents

Introduction

Agates are probably the most common gemstones on Earth. They have been used to create ornaments for around 7000 years, and have been recorded on every continent and in several different geological environments. Since the mid-1900s, agate collecting has become very popular in Europe, North America and the Middle East. There are thousands of agate-producing areas on Earth, and in the following pages, you will become acquainted with a small, but representative sample of these fascinating gems.

The formation of agates is of scientific interest to a range of scientists, from geologists investigating their geological significance to chemists studying the formation of chalcedony within water pipes in geothermal areas. This book is intended to present the current state of knowledge of these gems. Thus, the goal is to acquaint the novice and remind the advanced collector of the rich geological and cultural significance of these most beautiful stones, and to take the reader on a tour of the geographical distribution of agates across the world.

Opposite Moss agate found in the 19th century, from Tayport, Scotland.

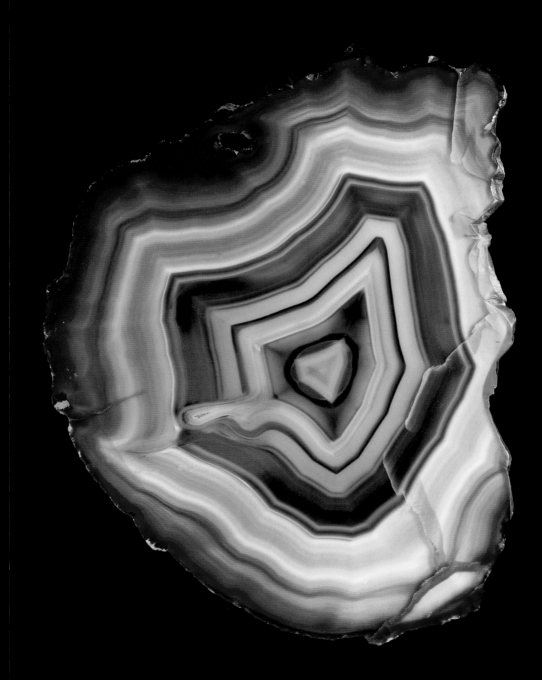

Names of agates

In the history of mankind's desire to collect examples of the surrounding natural world, agates, with their varied patterning and often vivid colours, must have been one of the first geological specimens to have been collected. It is therefore not surprising to find that many of the names we use today are also as ancient. Names such as chalcedony, agate and sard were used in the early years of the first millennium by people such as Pliny and Theophrastus. This was long before any kind of systematic naming was devised, and before even the science of mineralogy was born. These names are still used, and while not having rigid definitions (as modern-day descriptions would require), are useful in separating the major types, and have a general acceptance in the mineralogical world. Many new names have been invented via the popular literature to describe the many types of chalcedony, and some of these have become established among collectors. While they may be useful to allow collectors to communicate details of findings, they have no real scientific status. The mineralogical community does not define mineral varieties so while quartz (a species) is defined and has established physical and chemical characteristics, chalcedony is not defined but is accepted by common usage.

Mineralogical names

Among the cryptocrystalline varieties of quartz, are the chalcedonies and agates. Some authorities see agate as a variety of chalcedony, others as a near synonym. The name chalcedony goes back at least as far as Theophrastus and Pliny (AD77). Agate, is a term derived from the River Achates (now Drillo) in southwest Sicily, but the ancient word *achates* is not wholly synonymous with modern-day agate. Pliny distinguished many types of agate according to their colour, banding, fanciful resemblance to various objects (e.g. "eye(d) agate") etc. In 1747 Wallerius, a Swedish mineralogist, remarked that, "to enumerate all the varieties of agate is

Opposite Agate from Uruguay.

impossible and unnecessary", but it still continues, and seeing scenes in the patterns was as common then as it is today. Some authors of the period divided agate (known as *achates figurate* – agate with figures) into *achates anthropomorphos* (when they showed human forms), *achates zoomorphos* (when they showed animals) and *achates phytomorphous* (when they showed plants). Agate is today a truly all-encompassing name!

There are though other types of chalcedony, which are accepted by common usage. Iris or rainbow agate is the type which shows spectral colours due to the huge number of fine bands (maybe as many as 17,000 to the inch!). Sard (*sardion* of Theophrastus) has the distinction of having the oldest name known to definitely apply to a type of cryptocrystalline quartz, and is derived from Sardis, the capital of the ancient kingdom of Lydia (now in Turkey). Later, in the Middle Ages, the term was split into carneolous (now carnelian, from the Latin for fleshy), and cornelian, derived from the Latin (cornus) for the cornelian or dogwood cherry tree, which has a reddish berry. Moss agate is that agate which contains inclusions (usually green) which resemble moss. Other useable names are chrysoprase, plasma, prase, heliotrope, bloodstone, flint, chert, novaculite, and jasper. Simple descriptions of these will be found on pp. 39–43, but it should be noted that some of them are close to each other in appearance, and are not very distinguishable. It is sometimes arguable into which category a specimen actually belongs.

Names have some value in helping a collector determine something about the geographic locality and the geological setting from which an agate has been collected, but names can also be misleading and confusing, especially to the novice.

Geological names

Some named varieties of agates could be considered to have valid geological names. Such names result when a kind of agate is described in a scientific journal along with a record of its stratigraphic and geographic occurrence, e.g. Lake Superior agates that are common to the northern midcontinent and Great Lakes region of the USA and the Great Lakes region in southern Ontario, Canada. A geologist named A.C. Lane recorded these agates in a Michigan Geological Survey Bulletin in 1911 and used the name Lake Superior agate. Ellensburg blue agate (Thomson, 1961) of Washington in the northwest of the USA is another such example, and Queensland agates from Australia may also fall into this category, inasmuch as a geologist named Cameron (1900) included their occurrence in a gold mining prospectus.

Local names

Some agates carry local names for the region in which they are found. Parana agates (often mistakenly called Piranha agates) from Rio Grande do Sul, Brazil, are such an agate. Fairburn agates derive their name from a settlement called Fairburn, South Dakota, and Lucky Strike thunder eggs from Oregon derive their name from the mine that produces them. Ardownie Quarry agates derive their name from Ardownie Quarry in Scotland, and M.F. Heddle named agates for about 50 sites in Scotland. These names have become established by usage and tradition rather than by publication. The names may have appeared later in print, but they were well established long before publication.

Trade names

Some agates carry trade names that may be a registered trade mark of a gem producer. Such a name is royal purple Aztec agate from Chihuahua in northern Mexico. A recent addition to this list is Ocean Jasper® that contains a large percentage of banded agate in a spherulitic matrix. Such trade names are registered properties and can be used in sales only by the named owners or their authorised agents. The same kind of agate that is covered by a trade name may be found in another mine a short distance away from the mine that produces the agate with the protected name. Hence, you will often see variations of the protected name, for example, Aztec royal purple, for agates produced by these other nearby mines. Trade names have but a single purpose – to boost sales of the agate.

Trivial names

Trivial names are those given to a particular agate by a collector or dealer to honour some special occasion such as a family event, a business event or a celebration of some achievement. Such examples are the 'stopsign' agate which was found near a major road stop sign, and the 'ankle' agate that rolled from a cut and struck the finder on the ankle. Such names have appeared in the published literature, but they have no meaning outside of their immediate environment. This practice became more popular when an author of a popular publication stated that named specimens were worth more money than unnamed ones.

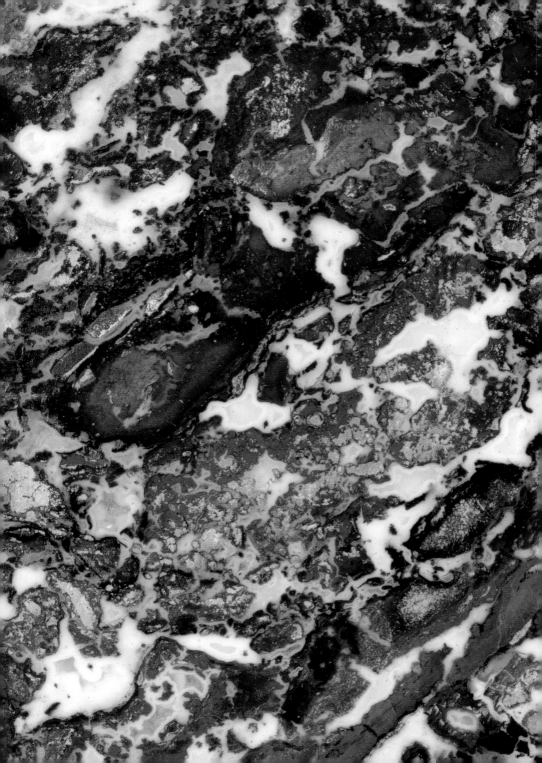

What is an agate?

Silicon dioxide in its crystalline form is called quartz, with the chemical formula of SiO_2. Chalcedony is a microcrystalline quartz made up of twisted crystal fibres. Agate is banded chalcedony, but agate may also contain the polymorphs crystalline quartz, moganite (which forms on the outside of the agate), and common opal. Agate has a microcrystalline structure composed of microscopically sized crystals. Most agates contain water (unlike quartz which is anhydrous), chemically bound with the silicon dioxide, but this amount is a small percentage and far less than in, say, common opal, which is a form of low temperature quartz, and has no recognised crystal structure (i.e. it is amorphous).

Formation

Agate occurs as a filling within gas cavities and veins in certain lavas or as a replacement mineral within some sedimentary limestones and claystones. It is most commonly found within volcanic lavas, in particular andesites, some rhyolites and tholeiitic basalts. The more explosive volcanoes produce ashes or tuffs – a mixture of ash, lava and sediments torn from the surrounding rock by the eruption.

All these varieties of lavas are very poor in free silicon dioxide, and the fact that agates are composed of silica has resulted in several theories being generated of where the missing silica originally came from. One of the most likely theories is that micro-shards of silica in glass explosively produced ashes and tuffs, devitrify and eventually release silica in the form of a watery gel that permeates through into underlying lavas via meteoric (atmospheric or rain) waters. Meteoric waters heated at depth, rise and dissolve silica *en route* to the surface. Other suggested sources of silica include decomposing vegetation or animal matter within sediments deposited between eruptions.

Opposite Jasp agate from Southeast Asia.

Above A vesicular lava showing empty gas chambers.

Above A typical amygdaloidal lava.

What is certain is that lavas contain gases, and that these gases may become trapped within the cooling and hardening rock, forming cavities called amygdales or vesicles. Such lava is called amygdaloidal lava because the original vesicular lava contained almond-shaped cavities (from the Latin 'amygdula', an almond). (In fact they can be totally misshapen, round, almost flat or most commonly bun shaped.) These subsequently become filled with agate forming materials. Agates can also occur in fissures within the rock called veins, or as long filaments, similar to but more numerous than veins, called stringers. More rarely, within sedimentary rock they form as nodules that are the result of the replacement of a former mineral, or former organic material like coral, which is more likely to become agatised than say sponge, as the latter are generally soft-bodied held together by a loose network of usually carbonate spicules that leave no real trace of a skeleton.

Although the formation of agate is surprisingly complicated, put simply, agate is derived from a silica-rich gel which fills the cavities. It is thought that silica-rich solutions either enter the cavities and then form gels or that gels form outside the cavity and move into the cavity. The gel is deposited in the cavities over a period of time, at a range of temperatures between 40 to 270°C, and it is generally thought to be in the lower range of between 50 and 60°C. Whatever the origin of the gel it is almost certain that the banding within the vesicles is triggered by chemical and physical reactions due to temperature and pressure within the cavity, and the degree of bound water held within the silicon dioxide.

It is important to remember that banding can occur as two distinct and not necessarily concordant parts: the so-called growth ring bands of chalcedony and those of coloured bands that are the result of the chemical deposition of, primarily, iron oxides. In the former case, silica gel is suspended in alkaline meteoric water or alkaline groundwater and is carried into the cavities in which the agate forms. When the silica gel comes into contact with incompatible chemicals (i.e. those with like electrical charges), it is theorised that these chemical impurities are rhythmically expelled from the leading edges of the microcrystalline crystals or fibres, so forming the visible growth bands within the nascent agate, the so-called Belousov-Zhabotinskii reaction. If the nodule (or vein) is entirely chalcedonic then the microcrystal growth is rhythmic, meaning successive bands form one after another, and are subject to the laws of spherulitic crystallisation (see below).

The gel is 'mobile' – in a jelly-like state – and subject to plastic deformation due to internal forces of crystallisation. These reactions cause the gel both to deform the banding and to sometimes expel some of the gel via tubes of escape (see p. 15). Given time, the gel crystallises from its nascent state from the outside in, by initially precipitating a thin 'skin' of often clear chalcedony followed by spherulitic

crystallisation – tiny, needle-shaped fibres of chalcedony grow abundantly in close and radial association with each other, forming a succession of minute growth bands. Small hemispheres (often called eyes but which are in fact half-eyes if the agate is cut in half) form under the skin also by this process. This in turn is followed by a more vigorous crystal growth where chalcedonic fibres grow at right angles to the surface, forming long and short intertwined fibres, the shorter ones forming banding due to the Belousov-Zhabotinskii reaction as explained previously.

Left *Polished volcanic rock showing vein agate.*

How agates form

Dense silica gel fills cavity skin.

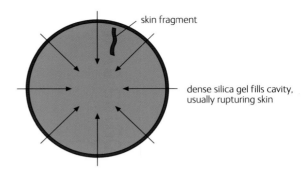

skin fragment

dense silica gel fills cavity, usually rupturing skin

Movement and separation of gel into hydrous and anhydrous layers begins.

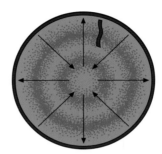

Separation continues and layering becomes more distinct.

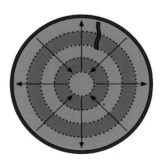

Clear chalcedony layer now crystallised.

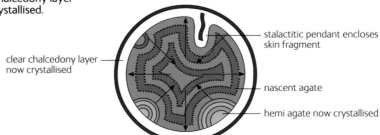

stalactitic pendant encloses skin fragment

clear chalcedony layer now crystallised

nascent agate

hemi agate now crystallised

Chalcedony fibres in clear chalcedony layer (spherulitic crystallisation).

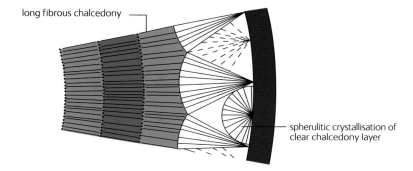

long fibrous chalcedony

spherulitic crystallisation of
clear chalcedony layer

Tube of escape with dilatation, rent in clear chalcedony layer, and quartz within crystallised layers of chalcedony.

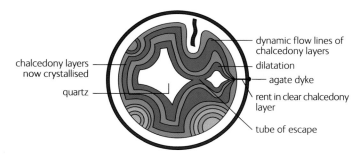

dynamic flow lines of
chalcedony layers

chalcedony layers
now crystallised

dilatation

agate dyke

quartz

rent in clear chalcedony
layer

tube of escape

Onyx agate (also showing fortification banding).

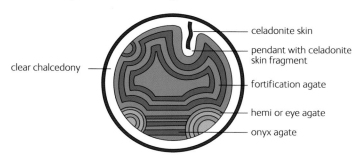

celadonite skin

pendant with celadonite
skin fragment

clear chalcedony

fortification agate

hemi or eye agate

onyx agate

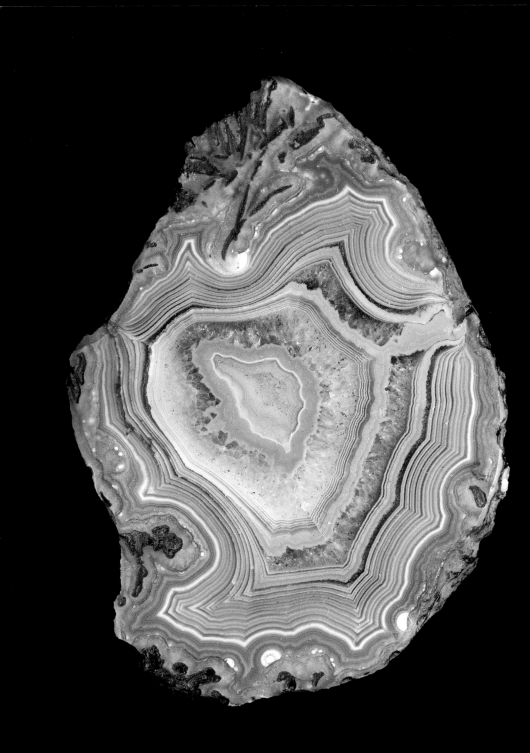

Because growth is at right-angles to the previous growth surface, the bands become progressively more contorted as the centre of the nodule is reached. The fibres never grow through or collide with the fibres opposite them even when there is insufficient space, so 'head-on collisions' are neatly prevented by so-called sector-zones, which form herring-bone patterns of banding, the point of contact being clear and visually fibreless. The pattern of banding is dependent on the degree of distortion of the gel, and the previously described growth.

One of the features commonly seen in wall-banded agates, but never in onyx are 'tubes of escape', which appear to be tubes running from the interior of the agate to the skin; in reality these are rents (tears or splits), caused by internal pressures within the nascent agate expelling some of the contents, and so forming a minutely raised ridge called an agate dyke on the outer surface of the agate. These tubes of escape have dilations that invariably appear just before the gel is expelled through the skin, possibly as a consequence of the gel having to break through the marginally more resistant skin and bunching up behind it. Subsequent crystallisation and banding then follows the 'tube', yet another reason why the bands do not often follow the contours of the vesicle. The reactions in the forming agate are sometimes sufficiently intense to cause water to boil within the dilation of the tube, and these can be seen in rare examples as little bubbles.

Many agates contain assemblages of chalcedony, crystalline quartz and other minerals such as calcite ($CaCO_3$) and silica-rich zeolites. In most agates the centre can be crystalline quartz which is anhydrous (containing no bound water). As well as forming the centre of agate nodules, the quartz also forms bands itself (of clear quartz) on which chalcedony can subsequently grow. The chemical conditions of the nascent gel and the physical conditions imposed by temperature and pressure result in loss or gain of bound waters, and thus the precipitation of quartz or chalcedony. If the concentration of the gel is low, there is less growth material available and consequently an agate with a hollow centre occurs. This is called a geode and sometimes contains clear or coloured quartz such as amethyst or smoky quartz. The geode is hollow, has well-formed pyramid-like crystal terminations or points, and if surrounded by banded agate is highly prized; such specimens are commonly called 'coconuts' in the USA. One recent theory on

Left A section of fortification agate from Scurdie Ness, Scotland showing the distinct banding.

Above *Blue lace agate from Transylvania, Romania, showing the hollow centre containing coloured quartz typical of the geodes.*

agate formation is that the crystallisation and resulting banding in the agate could be the product of the silica polymerising – forming long chains of identical molecules – as thought by Peter Heaney. He suggests that the initial deposition layer is moganite, which is a mixture of chalcedony and quartz, followed by pure chalcedony and quartz. Polymerisation is thought to occur as the silica precipitates in short, repeated units of five to ten molecules, and if the water content is high enough the silica polymerises. The banded structure of an agate is then formed when the silica/water content is high enough to give repeated polymerisation, with the fibrous crystals forming very rapidly. If water is no longer available then pure crystalline quartz forms, as is often seen nearer the centre of agates.

The outermost surface of the agate's skin is often covered by a soft, green mineral called celadonite, a decomposition product from lavas. This outermost skin is often green though red, brown and white are found as well. Most of these softer and very thin layers are composed of minerals of the chlorite group. The outer surface of celadonite is smooth, the inner surface is irregular due to shrinkage. The chalcedony forms a mould of this inner surface, thus when the celadonite is removed the outer surface appears pitted. Most agates throughout the world have these pits, and overall the general appearance of many agates once cut is that of a solidified jelly. Celadonite can be ruptured or torn apart by the infill of the gel, and form greenish filaments within the agate. This is called moss agate but does not have an organic origin. When present, certain minerals in the gel can combine to form other minerals such as calcite and members of the zeolite family. Subsequently, agate banding, shape, form and patterns, will be influenced by these earlier crystallisations. If they form at a later date, they often produce well-developed and beautiful crystals within geodes.

Many agates are a hotchpotch of the above minerals, sometimes emerging as wonderful specimens, other times ruined by them. Whilst there is a general depositional order this is not constant. The order is: celadonite, zeolites/calcite, clear chalcedony, banded agate. No one has yet understood the extraordinary conditions found within a nascent agate, and why certain minerals precipitate rather than others. Because these conditions will alter radically within the rock, agates are often very different even from their near neighbour in the next-door vesicle. It is all unclear and even though research continues on simple agates composed of chalcedony, an understanding of the broader picture within the rock is still a long way off.

Reduced to its simplest terms, there are only two essentially different kinds of banded agates: wall-banded agates where there are growth rings following the approximate outline of the cavity, and level agates, normally called onyx, where the bands form gravity-controlled horizontal parallel layers. This latter variety is less fibrous and more granular in crystal structure and possibly has a higher bound-water content. It is also much less dense and compact than wall-banded agate, and for a variety of reasons fails to adhere to the sides of the cavity, and is gravitationally deposited in flat layers. Commonly, the onyx layers form at the base of the cavity, whereas the dome or top section of the cavity forms wall-banded agate, which can be of an entirely different colour. Other wall-banded agates contain internal sections of onyx, where the precipitating gel ceases to adhere to the walls of the vesicle, but becomes subject to gravitational layering, probably due to an increase in the amount of bound water within the gel.

Agate then is a mineral with a microcrystalline structure, and a hardness of between 6.5 and 7 on the Mohs scale of hardness (see p. 45), which means it is hard enough to scratch glass. The internal crystal structure is arranged as compact and twisted microscopic fibres, always oriented to their direction of growth. Some agates are more compact than others, and on the whole the best quality material is of this type. Less compact agate tends to have less of a microcrystalline fibrous

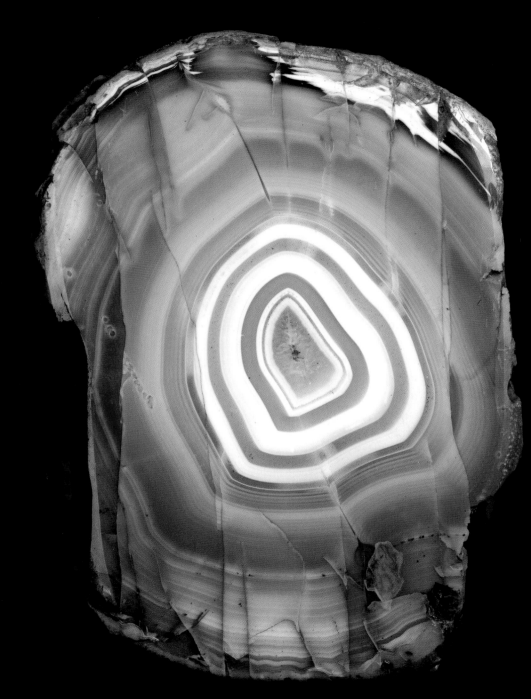

structure and more of a granular one, although it is common for many kinds of agate to shade imperceptibly from one to another. The most desirable agates are banded, often with several colours and composed of many bands of varying thickness. Where the lavas contain impurities such as iron, this results in more vivid and variable agate colours. Many agates have internal cavities which contain crystalline quartz, which forms as clear, translucent six-sided crystals.

Colour

Pure chalcedony is clear, honey coloured or grey. It is semi-opaque to, rarely, translucent unless cut into thin slices. However, most agates are red and blue, although in reality the 'red' will vary from pale pink through orange to pillar-box red, whilst the 'blue' will vary from grey-blue to cornflower blue through to almost black. Other, rarer colours include yellow and green, or white bands standing out from the background hues. Dark browns, even blacks and combinations of all of the above can produce a whole range of strong to pastel shades, each agate being either subtly or completely different from its neighbour.

Almost all of these great varieties of colour are due to oxidised iron. Although shades of colour may show a preference for certain areas within the agate or even preferentially around some bands, colour by itself has nothing to do with the growth of an agate, but is a by-product of iron-rich solutes entering the gel either initially or subsequently after the agate has crystallised. The one exception to this is the formation of some pure white to whitish-blue bands that are of an amorphous opaline mineral. Such attractive and often thick bands are common in onyx, and a particular feature of some Scottish agates. Close examination of a strongly coloured red or yellow agate with a hand lens will show that, in many cases, the colour is present in the form of tiny spheres, suspended in a clear chalcedony, the red composed of haematite and the yellow of goethite. The red spheres are sometimes called haemachatae.

A curious and noticeable feature in some localities is that the agates from ploughed fields differ in colour to those from nearby rock. This is probably due to intense weathering producing strong iron compounds that infiltrate the agate over a geologically short period.

Opposite Agate from Oberstein, Rhineland, Germany
showing the growth rings following the cavity outline.

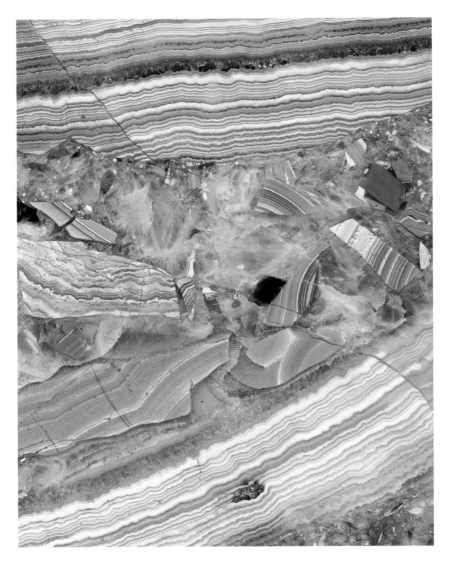

Above Brecciated or faulted agate showing the red and
yellow bands of haematite and goethite.

Many cracked agates show preferential colour staining, often red or pink, that follow crack lines. Indeed chalcedony has a strong ability to absorb iron-rich groundwater at any stage after its formation and this can lead to attractive specimens. As a gross simplification, agates tend to form either 'blues' or 'reds' with a bias towards one or the other depending on the locality, and with any shade in-between plus other rarer colours.

Cracks

A significant percentage of agates once cut will exhibit cracks because chalcedony is very brittle. Agates are millions of years old and have been subject to geological stresses and strains, and if near the surface in weathered rock are often further subject to erosive forces as well. Agates found on the surface are also subject to frost and heat shattering, not to mention the action of ploughs in fields. The best agates come from rock which has changed very little, whilst the most numerous and easy to collect (but most cracked) come from ploughed fields.

Occasionally, cracks within the agate have been re-cemented by calcite or even chalcedony, proving that many cracks are extremely old. As a rule of thumb one in ten agates once cut is worth polishing for a cabinet specimen. However, they may provide rough material for smaller finished gems. When broken, agates tend to produce a series of concentric grooves which was initially thought to resemble the growth patterns of some shells, and so is called 'conchoidal' (from the Latin 'concha', a shell), although this is more pronounced in non-banded chalcedony.

Agate terminology

The terminology used to describe agates is confusing, mainly because many of the terms are jewellers' terms, not mineralogical ones. Since agate by its nature is the chameleon of the mineral world, new names based on visual appearance keep being invented, when some of the traditional terms adequately describe these variations. Some of the more fancy names are used as a means to stimulate sales and make the agate sound attractive to potential buyers. Furthermore many agates exhibit features that incorporate several of the following nomenclatures, only confusing the issue further. That said, the following should give the reader a reasonable grasp of the varieties.

Wall-banded agates

Wall-banding is simply a term to describe the bands that follow the shape of the vesicle. It is a factual term, unlike most others, which are subjective or commercial, or both.

Agatised coral: Agate and chalcedony can replace the calcium carbonate of marine organisms, filling the internal voids. Where agate is found in sedimentary regions (particularly Florida in the USA), a siliceous gel can engulf the shell of an already dead animal. Several species of marine organisms have been 'agatised', namely corals, bryozoans, crinoids, clams, ammonites and trilobites.

Agatised wood: This is a rather special type in that it is not really agate at all but chalcedony that has permeated through the internal wood cell area. Often the form of the cell walls remain, and although appearing quite tree-like, it is in fact stone. As chalcedony has a range of deposition temperatures it can replace organic remains provided that a source of silica is available. Other areas of deposition include sinter or hydrothermal deposits around hot springs, chalcedonic deposits in certain rivers and within water pipes in areas of hydrothermal power.

Brecciated or ruin agate: This is agate that has been subject to geological pressure and has had its original banding structure fractured and re-cemented by chalcedony, agate or other minerals to resemble ruins or angular fragments. True

Below *Ruin agates are formed when a pre-existing agate is shattered and the fragments are later re-cemented by new influxes of silica. Specimen approximately 5 in (12.6 cm) wide.*

ruin agate has possibly been subject to shear forces within a fault and appears somewhat more like building rubble than brecciated agate which has a more random appearance.

Calico agate: This is an American term that refers to small patterns of banded agate set amidst a groundmass of a lighter colour.

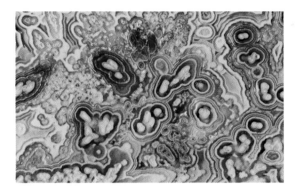

Left Close-up of the cut face of agate stone showing marble-like patterns. Unknown locality.

Dendritic agate: This has dark brown to black inclusions formed by minerals containing iron or manganese oxides. Dendrites are possibly made up of colloidal or minute-sized particles that are suspended in water and gain entry via osmosis. They are commonly confined to the area between separate bands within the agate producing a moss or fern-like structure.

Right Large cabochon of dendritic agate from Snake River area in Idaho. Width approximately 7.5 cm (3 in).

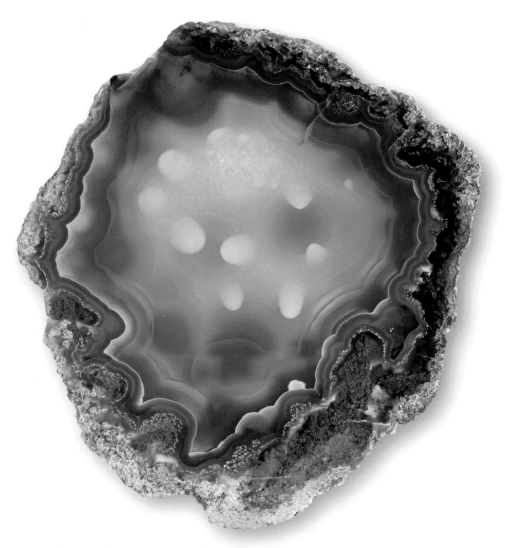

Above *Agua Nueva agates are named for Agua Nueva Ranch, in northern Chihuahua, Mexico. They are usually mined from relatively unweathered rock and the surfaces of the nodules show many features that are lost on more weathered agates. Note the white, disc-like inclusions. Their bands are generally sharp and brilliantly coloured and many contain mossy inclusions. Height of specimen approximately 15 cm (6 in).*

Disc-bearing and ovoid–bearing agate: An unusual variety wherein the chalcedony contains small white discs or egg-like ovoids within the layers. These structures are now shown to be mixtures of chalcedony and opal.

Eye agate: There are several types of structures within an agate that produce 'eyes'. The 'eyes' can be formed from slicing through at right angles to the length of a stalactitic chalcedony formation and by cutting banded hemispheres parallel to the base. The stalactitic formations are the result of chalcedonic crystallisation around a needle-like crystal, often of a type of zeolite that may well have formed within the cavity before the entry of the siliceous gel.

Below The classic eye agate is one in which hemispheres of agate are situated with their bases toward the outside of the nodule. Specimen approximately 5 cm (2 in) long.

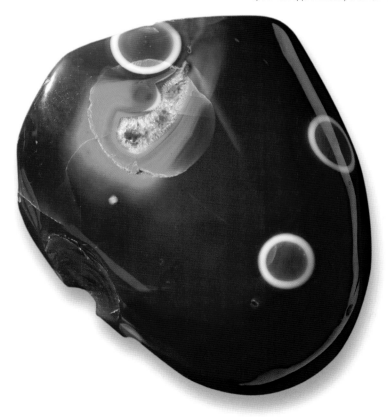

Above Numerous faults can be seen in this agate from Fintry, Stirlingshire, Scotland.

Faulted agate: Occasionally an agate will be faulted and re-cemented into a slightly different position due to rock movement. The rents are closely joined, usually by chalcedony or calcite, and each 'segment' of the agate can be of a different colour, stained by the introduction of iron-rich waters into the rents. Mostly these are in the form of blue agates or onyx that have reddish or yellowish colour zones either side of the rent.

Flame agate: This is an even more subjective name than most. In Europe it refers to any banded agate that has a flame-like banding within the agate, normally of a red colour, whilst in North America it refers to stringers of any red mineral, probably haematite, that form within a clear chalcedonic matrix, giving the appearance of a forest fire.

Left In Europe, flame agates are considered to be those whose bands resemble a candle flame. This specimen from Malawi, Africa is approximately 6.3 cm (2¹/₂ in) high.

Below Fortification agate from Burn Anne, Scotland.

Fortification agate: This is a common type and refers to any wall-banded agate that has sharply defined banding caused by the fibre orientation. In plan view it appears rather like a medieval fortification.

Iris or rainbow agate: This agate is usually fine-banded and rather colourless. Although seemingly dull, when thinly cut and held to a source of transmitted light, the fine bands act as a diffraction grating, breaking up the beam of white light into its spectral colours. This iris effect is not confined to poorly coloured agates, but can also occur in brightly coloured ones too. However, if the agate is of a strong red for example, the red will be masked by transmitted light, leaving only the blue and yellow spectral colours to appear.

Below Iris agates are those agates that have sufficiently fine bands or crystal structure to act as a diffraction grating and break light into its component colours, red, yellow, and blue. Specimen from Rancho Coyamito, Chihuahua, Mexico, approximately 5 cm (2 in) square.

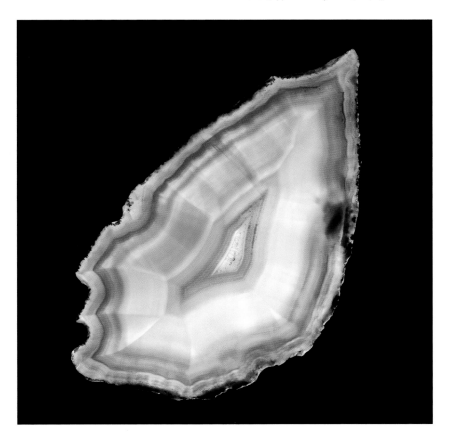

Jasp agate: Chalcedony that is admixed with other oxidised iron-rich minerals forms a common variety of chalcedony called jasper. It is usually red, yellow or brown and if the siliceous gel contains a mixture of agate and jasper, a variety of agate called jasp agate is formed. It is also usually red or yellow, often opaque except in thin edges and usually beautifully banded as the strong colours are clearly defined.

Left *Jasp agate specimen, unknown locality.*

Below *Lace agate is also a generic name as this kind of agate is found in several locations throughout the world. Most lace agates form in veins rather than as nodules. Length of slice approximately 25.4 cm (10 in).*

Lace agate: This is often found in veins as well as nodules and classically it is finely banded blue. The sharp herring-bone patterns make it appear like finely wrought lace. It is common to have the pattern interrupted by pseudomorphs or false forms of other crystals such as calcite or zeolites, that have been replaced by chalcedony in the original shape of the former crystal.

Above Mocha agate, unknown locality.

Mocha agate: This is a type of dendritic agate where the moss-like patterns are predisposed to follow cracks as well as infiltrating between growth bands, giving arborescent forms.

Moss agate: If dendrites or small branch-like structures of chloritic material on the outer skin of the agate are forced in with the silica gel and are green, then they are formed of celadonite (or if brown then some other related mineral of the chlorite family). Geologists used to think that this was petrified moss, but it is inorganic and can form either delicate dendritic branching structures, especially in some of the Indian localities, or else a mixed up mass of 'moss' and chalcedony with sometimes attractive agate banding around the moss.

Above Moss agate, unknown locality.

Plume agate: This is essentially another type of moss agate but in this case the plumes form arborescent structures, often of great beauty, within the agate. Plumes are often bright red, usually of a haematite, or bright yellow from goethite, one of the products of weathered haematite.

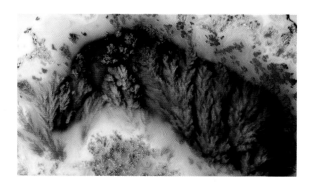

Right Moss or plume agate, unknown locality .

Above This sagenitic agate from Dunbog, Scotland, shows a spray of fine chalcedony tubes which would have formerly enclosed needle-shaped crystals, most of which have long since disappeared.

Sagenitic agate: This catch-all term refers to any agate that has needle-like crystals within the chalcedony, notably rutile, goethite, anhydrite, aragonite or a member of the zeolite group. There is no such mineral as sagenite. Chalcedony does not always form tubes around the intrusions so leaving them clearly visible. Often, and particularly in the case of zeolites, the needles appear in the form of a spray with the agate banding forming intricate contortions around them. The needles can be replaced at a later date by chalcedony forming a pseudomorph.

Stalactitic agate: Some localities contain agates that have pendulous stalactite formations of either banded or non-banded chalcedony that form in voids and more commonly later are enclosed within the agate. These interesting structures either form around a sagenitic crystal or else accrete around a slight roughness or protrusion in the cavity wall of the nascent agate, forming mammilations. Not all stalactites grow vertically downwards; in some localities they are bent or even twisted, although these curiosities are found invariably in voids. Some stalactitic inclusions in agates resemble the stalk aggregates described by the Russian geologist Lev Lebedev. These stalk aggregates form when a particle of lower density rises upward through a liquid or gelatinous medium of a higher density, causing them to be uniform in size and length, whereas stalactites are commonly thicker near their tops than near their tips.

Tube agate: If the agate is cut parallel to the tube or stalactitic formation the agate is then called a tube agate.

Right Tube agates form when needle-like crystals of a mineral such as goethite provides the template around which the agate grows. If the cut is made perpendicular to the crystal, an eye-like structure appears, such as in this Coyamito agate from Chihuahua, Mexico. Diameter of specimen approximately 5 cm (2 in).

Level-banded agates

Landscape or scenic onyx: This is a catch-all phrase describing any onyx that looks like something recognisable. Ordinarily these form 'sea and sky' agates, complete with beach, waves and sky (with a few clouds). Any number of permutations on this theme are locally possible depending on the locality. It is common in some areas to have an agate that has a base of onyx and the topmost part or dome made out of wall-banded agate. These are called onyx-agates but are a composite of both types rather than another separate type. Occasionally highly complex agates containing different layers of agate and onyx occur.

Below This agate is from Espumoso, Brazil, a region famed for its landscape agates.

Onyx: Level-banded agates are traditionally called onyx after the Greek word for the human nail, which is horizontally ridged when viewed side-on. In some areas onyx is called waterline agate, and Michael Landmesser in Germany uses the term 'Uruguay' structure. Here, the banding alternates, usually between blue and white or even on rare occasions, black and white, the white often being made of cachalong, an opaline-rich variety of chalcedony, which is more granular in structure than fibrous chalcedony. There are several different varieties of onyx in common usage, as described below.

Plynthoid onyx: Occasionally, the horizontal bands that make up the layers rend vertically rather than horizontally, possibly due to shrinkage, allowing the component parts of the layer to form brick-shaped accretions. The intervening spaces, if narrow, fill with chalcedony and if wide, often with crystalline quartz or, more rarely, with small independent agates.

Sard onyx: Here, the banding is of a red to brown colour, either alternating with white or with blue.

Above Sard onyx, unknown locality.

Above Thunder egg, unknown locality.

Thunder eggs: The term thunder egg and several variants are common to several cultures and the exact origin of this curious name may never be determined. They are a type of unusual agate only found in the more explosively produced ashes and tuffs associated with rhyolites and perlites. This type of agate is encased in an outer shell, usually dark brown in colour, of the rocky matrix mixed with silica glass. Within this shell is the actual agate, star-shaped due to pressures within the plastic ash, and containing onyx, often with a dome of wall-banded agate. Some researchers consider thunder eggs to be the first agates to form in the agate cycle, being stratigraphically young, and that if the thunder eggs weather-away, their silica is used to fill empty vesicles below in underlying lavas, thus producing the agate equivalent of 'new wine in old bottles'. Although this theory is attractive, considering how resistant agate is to weathering and that Precambrian agates (over 545 million years old) from the Great Lakes area of North America are still very much intact, it does rather suppose that the agate and onyx within the eggs are more susceptible to weathering than other agate types.

Wave onyx: In this instance, some but not all of the horizontal layers form hummocky curvilinear forms possibly due to an initial roughness or impurity that makes for repeated domed forms. It is very attractive and not that common.

Other types of chalcedony

Not all chalcedony is banded. Sometimes it is clear, dark, brightly coloured or mixed with impurities so that it is given another set of names, accepted into common usage. It is quite common to find faintly banded or non-banded varieties of chalcedony with agate in the same locality, and where agate starts and ends is partly subjective, but it is all chalcedony.

Bloodstone: Possibly the most attractive of the 'green' chalcedonies, this is a generally opaque dark green colour with blood-red spots distributed through the matrix. In some areas (notably the Island of Rhum off the west coast of Scotland), bloodstone is incorporated into agates proving, once again, that all the chalcedonies of whatever colour and other differences can meld into any number of entertaining sub-species of agate.

Right Bloodstone is characterised by having dark red spots in a dark green matrix. Bloodstone may grade into plasma as the oxidation state of the iron becomes lower. Specimen is approximately 5 x 7.5 cm (2 x 3 in).

Above The Rio Grande do Sul province in Brazil has provided some outstanding examples of carnelian agate, as exemplified by this polished slice that is approximately 12.5 cm (5 in) long.

Carnelian: This is flesh-red, translucent chalcedony, somewhat variable in colour. It is often extremely brittle and nodules of carnelian, although sometimes large, are frequently broken up by myriads of cracks. If an agate contains carnelian, then it is usually banded but tends to suffer cracks more than other varieties. Why it should be more brittle than banded agate or other forms of non-banded chalcedony is unclear.

Chert: This is another impure form of chalcedony, this time admixed with clay, and the resulting stone is white to yellowish. It too has been used as a building material and breaks with a conchoidal fracture.

Chrysoprase: This is a more attractive variety of prase, which is of a light green to apple green, translucent to opaque material. These two varieties of chalcedony are separated more by nomenclature and fashion for a colour than by any mineralogical differences. The green colour of chrysoprase is considered to be due to included, fine-grained, nickel compounds.

Above This piece of rough chrysoprase is from New Caledonia and measures approximately 2.5 x 5 x 5 cm (1 x 2 x 2 in).

Right Cabochon of chrysoprase from Australia. Length approximately 3.2 cm (1¹/₂ in).

Flint: This is an impure form of chalcedony, formed from a gel and found within sedimentary rocks, in particular chalk. Southern England is famous for its chalk deposits and the nodules of flint found within. These nodules are supposed to be replacements for types of sponges and are purple-black to brownish in colour with white sections. It has a fetid odour when struck due to the amount of the mineral pyrite and creates sparks. All flint is opaque, even on thin edges and, like all chalcedony, has a clam shell-like or conchoidal fracture. Flint was therefore highly prized as a source of fire from prehistoric times to the mid 19th century, when the safety match was invented. It was also the raw material for tools used by early humans, as the stone is capable of being knapped into useful cutting shapes. It is still used as a building material and there are several different varieties of flint-rich stone, such as the puddingstone from the English county of Hertfordshire.

Jasper: The mineral jasper has already been described under jasp agate where it forms a sort of halfway house between both types of chalcedony. Jasper by itself is an opaque form of chalcedony with admixtures of clays, and is coloured by haematite to give red and goethite to give yellow or brown. It can also be green on occasions, possibly due to chrome oxide. These oxides commonly colour all the chalcedonies, as mentioned previously, but jasper differs from agate in that it forms massive pieces in veins as well as nodules, sometimes in pieces weighing tonnes, and is not banded as such but the colours within can form 'swirls' of contrasting pattern and hue.

Right Cabochon of jasper from Banda,
Bandalkhand, central India.

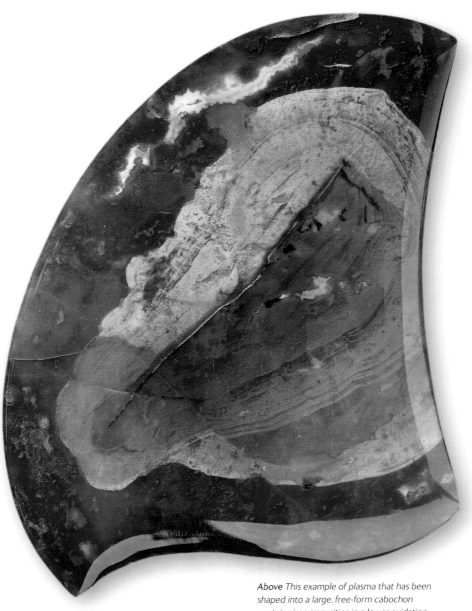

Above This example of plasma that has been shaped into a large, free-form cabochon contains iron impurities in a lower oxidation state than bloodstone. The stone is approximately 5 x 7.5 cm (2 x 3 in).

Plasma: This is generally opaque dark green with yellow or white spots, the colour attributed to celadonite or other members of the chlorite group of minerals.

Prase: This is a yellow-green to leek-green variety of chalcedony. It is translucent and non-banded and also quite rare.

Sard: This is orange-brown to brown chalcedony that is not as red or quite so translucent as carnelian.

Left Prase is typically a light yellow-green or leek green granular form of silica.

Below This slice of sard from Brazil is of a lighter shade than the carnelian and measures approximately 12.5 x 15 cm (5 x 6 in).

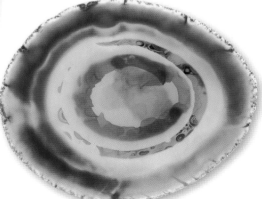

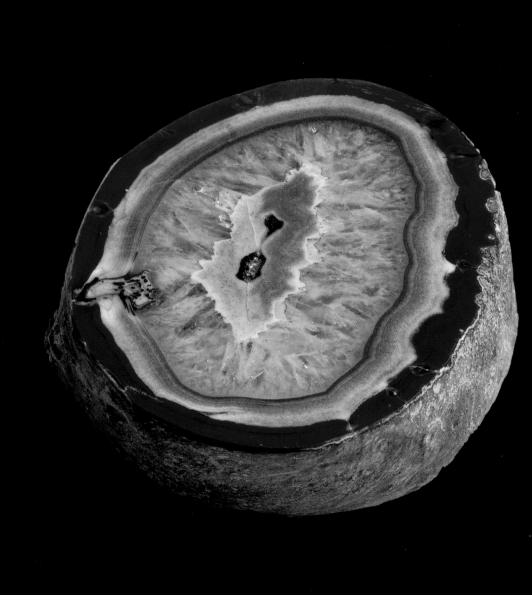

Properties of agates

Agate and all of the above named varieties are composed of silicon dioxide, SiO_2, one of the most abundant minerals on the surface of the Earth. Silicon dioxide is most familiar to us in the form of sand, which crystallises in the hexagonal system. Silicon dioxide, however, is not limited to crystallising in this system, and its many varieties can form crystals in several of the other different crystal systems (polymorphism). Wall-banded agates are made up of microcrystals of chalcedony that are arranged in twisted fibres, more or less perpendicular to the bands, and which X-ray data show to be of the trigonal class of the hexagonal crystal system (note that the trigonal class is only recognised by geologists and mineralogists in the USA as it is considered a separate crystal system outside the USA). Many moss agates and plume agates are granular and react as aggregates in polarised light, i.e. they remain light as a thin slice is rotated in a polariscope.

Hardness

Hardness is measured on a scale that the Austrian mineralogist Fredreich Mohs developed in the 19th century. It is an arbitrary scale based on the fact that each mineral will scratch those of a lower hardness.

The hardness of agate varies slightly from one kind to another dependent on its crystalline structure and the nature of the silica dioxide present. Chalcedony is usually taken as ranging from a low of about 6 to a high of slightly over 7 on a scale of 10. Some types of agate have been stated to be as hard as 9 on the Mohs scale, but in reality, they can be scratched with topaz.

Opposite Jasp agate from Galenberg, Germany.

Mohs hardness scale

1 Talc, soft and greasy

2 Gypsum, scratched with fingernail

3 Calcite, scratches a copper coin

4 Fluorite, scratches calcite

5 Apatite, scratches most steel
   knife blades

6 Orthoclase (feldspar), scratches
   glass with difficulty

7 Quartz, scratches glass easily

8 Topaz, scratches quartz

9 Corundum, scratches topaz

10 Diamond, scratches corundum,
   silicon carbide, itself

Hardness in relation to common products

2.5	Fingernail
2.53	Gold, Silver
3	Copper penny
4-4.5	Platinum
4-5	Iron
5.5	Knife blade
6-7	Glass
6.5	Pyrite
7+	Hardened steel file

Toughness

Toughness is a measure of the resistance of a stone to being cut with lapidary equipment and its resistance to fracturing. Agates can vary considerably in toughness. The toughness is due mainly to the fibrous nature of the microcrystals that make up the bands. This toughness has sometimes been confused with hardness. Most non-banded, granular moss agates are less tough than many of the banded agates.

Fracture

Fracture is the way in which material breaks, and is visible in the nature of the surface. Very fine-grained, non-banded agates may have a conchoidal or shell-like fracture, whereas non-banded agates with coarser grains may show finely granular fractures. Banded agates will usually show a very fine, splintery fracture to a somewhat hackly (jagged) fracture, depending on the sizes of the crystals making up the bands.

Optical properties

The speed of light through different materials varies and the refractive index (RI) of a material is a measure of the speed of light through that material compared with the speed of light in a vacuum. For agates the RI varies from 1.52 to 1.54 (water is taken as 1.33 and diamond is 2.42), which is very low. Agates have either fibrous or granular structures and this affects the optical character. Granular agates in polished sections show no variation and remain light when rotated in a polariscope, whilst fibrous ones show alternate lightening and darkening of fibres.

Above Polarised light micrograph of a thin section through a geode. Layers of chalcedony surround an interior of quartz crystal. Central hollow approximately 2 cm (0.8 in).

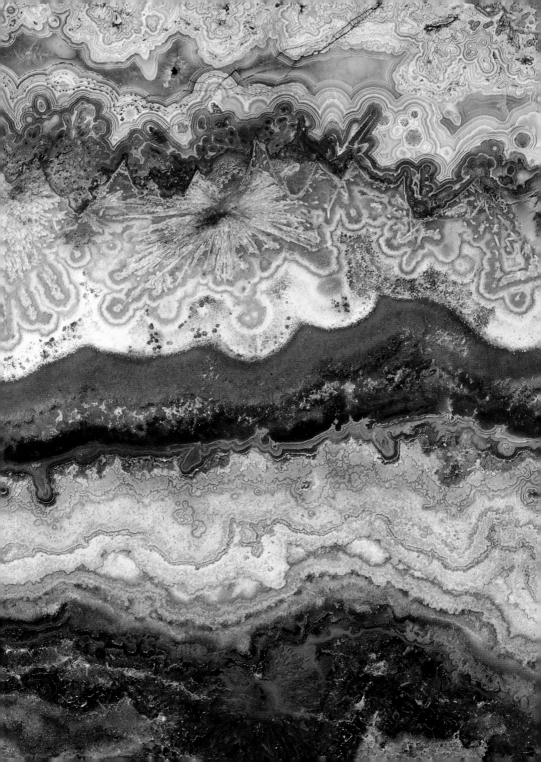

Sources of agates

Agate has been found on every continent and sometimes in abundance. Although an area may be renowned for its agates, only a small percentage will actually be of gem (or display) quality, having suitable colour, pattern and freedom from unsightly flaws and stains that it could be worn or displayed.

Geological occurrences of agate are considered to be equally as important as geographic occurrences, and listed below are six different kinds of geological environments in which agates have been found:

1   From explosive eruptions of rhyolitic volcanic rocks. Welded ash-flow tuffs are rhyolitic rocks with abundant free quartz in which the glassy shards are fused together. The magma which formed these rocks has the consistency of peanut butter. When these rocks are extruded to the surface, they are accompanied by violent explosions that may destroy all plants and animals over several hundreds of square kilometres of terrain. The kind of agate these rocks produce are called thunder eggs (see p. 38), and they have an irregular geometry, often with a star-shaped interior. For example, the Mount St Helens eruption in Washington, USA in 1980, may eventually result in thunder eggs and irregular to star-shaped agates.

Opposite Crazy lace agate from Mexico.

Right Thunder eggs generally have geometric to star-shaped outlines as well as a shell that is composed of rhyolitic or quartz-rich volcanic ash that contains numerous ash shards. Diameter of specimen from Friends Ranch, Oregon, approximately 12.5 cm (5 in).

2   From less-explosive basaltic or andesitic lava flows. Tholeiitic basalts are ultrabasic rocks that have little, if any, free quartz, and they are derived from a magma that has much lower viscosity than that which forms ash-flow tuffs. Amygdaloidal (almond-shaped) agates (see p. 12) are formed in these rocks. Some phases of basaltic or andesitic eruptions can produce large amounts of ash.

3   Formed in thick limestone beds that were deposited in relatively shallow seas creating marine sedimentary agates. Marine-deposited shale beds do not commonly contain agate nodules, but there are a few such occurrences. The cavities in the host rock may be solution cavities or burrows made by organisms such as worms, clams or arthropods.

Right Marine sedimentary agates form in limestone that has usually been deposited in near-shore environments. The cavity in which the agate formed may have been either a burrow made by some marine organism, the hollow shell of some marine organism, or a fracture plane in the bedrock. Specimen approximately 7.6 cm (3 in) in diameter.

4   Formed in wind-deposited siltstone and claystone creating continental sedimentary agates. The cavities may be formed from solution, tree roots or animal burrows, or they may be fault or joint planes.

5   From faults or fissures. Vein agates may form along the walls of faults or joints or other fissures in any of the above kinds of rock. If these agates are found out of place, it may be impossible to tell the kind of rock in which they formed, unless some of the matrix is adhering to the stones.

6   From ocean-floor spreading. Lastly, another kind of agate that has been more recently recorded in the geological literature has been observed in oceanic basalts that have flowed from volcanic rents along which the ocean floor is spreading. Such agates are very rare occurrences indeed, and even if they were common, the cost of their recovery would be prohibitive.

Right Vein agates are often characterised by having a layer of quartz crystals covering the outermost band of the agate, and the wall side of the agate is commonly very straight. Length of specimen approximately 30 cm (12 in).

All of the above scenarios require one thing: a source of silica. The author and A. Zarins suggested in 1993 that most silica sources for extensive agate deposition were volcanic and probably the result of large ash flows. They did not discount organic silica such as sponges and diatoms, but this source is of much less importance, even for agates that formed in marine sedimentary environments.

Worldwide agate occurences

The map below shows the locations of some of the world's leading commercial agate deposits as well as several that are of more geological or academic importance.

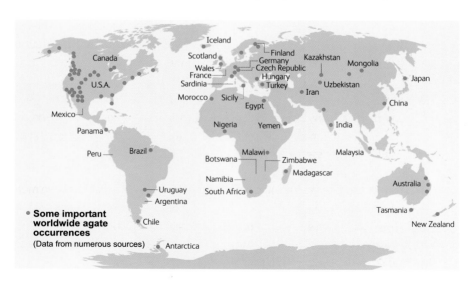

Europe

Sicily

Sicily should be considered the type locality for agates inasmuch as these stones were first described from there by the Greek scholar Theophrastus in 350BC. The word 'agate' was derived from the River Achates (now the Dirillo) that drains into the Mediterranean along the southeast coast of Sicily. A small city, Acate, that bore the name Achate in ancient Grecian times, is still extant along the River Dirillo, about 25 km (15 miles) inland. The southern part of Sicily, which is drained by the Dirillo, is composed largely of basaltic lavas with mixed pyroclastic rocks such as ash-flow tuffs of Pliocene age (5.3-1.5 million years ago). This is the ideal geological environment in which agates can readily form. Many of these Pliocene age basaltic and pyroclastic rocks were subsequently eroded, transported, and re-deposited as land-derived marine sedimentary rocks of middle and upper Pliocene and Pleistocene ages (3.5-0.5 million years ago) that are exposed in the Dirillo drainage basin.

Germany

Germany is one of the best-known historic sources of agate. Millions of beautiful stones have been mined from volcanic rocks of Permian age, and finished in the twin cities of Idar and Oberstein on either side of the Nahe River. The areas around Idar and Oberstein were possibly known agate producers even in Roman times, but they do not seem to have been exploited until about 1375.

Below *Agate from Oberstein, Rhineland, Germany.*

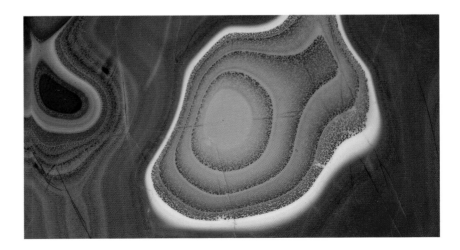

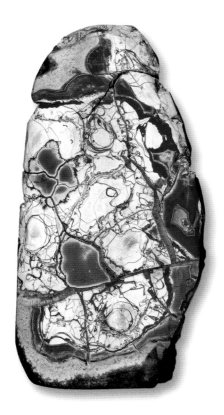

Above *Brecciated agate from Germany.*

Left *Jasp agate from Germany.*

The agates from the Idar-Oberstein area have been mined from volcanic rocks of Permian age (248-290 million years ago). This is about the same age as rocks from which the agate deposits at St. Egidien in Germany and Silesia in Poland have been mined, part of a rich agate-bearing terrain that extends from Thuringia and Saxony in eastern Germany through Bohemia in the northwestern part of the Czech Republic into Silesia in the southwestern part of Poland. Most of the agates from Idar-Oberstein appear to be amygdaloidal agates, whereas the agates from the latter two sites are thunder eggs.

Some of the agates from Idar-Oberstein have been recovered by surface collecting, but most have been produced from underground mines and tunnels. Maps of the Idar-Oberstein area show that over 50 mines have been active in the past. Of these historic mines, at least one near Steinkaulenberg is open to the public for tours on a fee basis. The operation is seasonal and literature provided by the mine suggests that visits should be planned ahead of time to avoid overbooking.

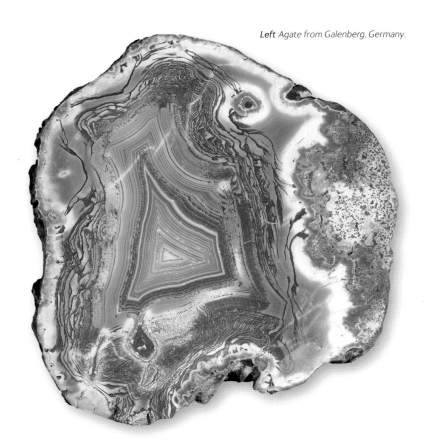

Left *Agate from Galenberg, Germany.*

As in most agate producing areas of the world, not every rock unit in the Idar-Oberstein area is a prolific source of agates. Many of the agates from this area are very brilliantly coloured; they rival the Mexican agates not only in colours, but also in contrast, structure, and freedom of flaws. Shades of red and pink are common, but practically every other colour common to agates have been seen in these specimens. Museum collections of agates from the Idar-Oberstein area include wall-banded and level-banded agates, moss agates, plume agates, eye and tube agates, and sagenitic agates to mention but a few varieties.

Many of the finest agates from Idar-Oberstein were collected in the 1600s and 1700s. That many of these agates were never turned into jewellery or *objets d'art*, is a testament to their natural beauty. Except for polishing a face on the agate, many were kept in their natural state, and this suggests that collecting of agate specimens has a long history in Germany and Europe. Even though many of the very fine agates were collected here several hundreds of years ago, some of the most beautiful examples of Idar-Oberstein agates have been collected since the Second World War and some even much later.

In addition to abundant supplies of agate, Idar and Oberstein were both situated on fast-moving, narrow streams that provided ample power for the development of an agate-cutting industry. The industry there survived on its own supply of agates before local deposits became somewhat depleted. The water power that was available in the Idar and Oberstein areas turned large sandstone grinding wheels that were up to 3.3 m (10 ft) in diameter. The vertical wheels did the rough grinding, then the finer sanding and polishing was carried out on a lap using felt or leather discs charged with a polishing compound such as stannic (tin) oxide. Skilled workers were capable of working on eight stones at a time. They lay on their stomachs and held stones attached to a dop stick between each finger, and held the agates to the revolving wheels until the required degree of smoothness and shape had been acquired. All this was incredibly unhealthy and ingestion of silica particles must have led to silicosis. The discovery of abundant agates in Uruguay and in Rio Grande do Sul, Brazil, by German emigrants to those areas revived the faltering agate cutting industry in Germany, and now much of the agate worked there is from the Uruguayan and Brazilian sources.

Above *Old-time method of gem-cutting in Idar-Oberstein, Germany.*

Czech Republic

The area around Horny-Halze, a small village in the northwestern part of the Czech Republic near its border with Germany has produced agates for many years. Fine examples from here are in many museum displays throughout the world, and many of these stones now grace private collections. Although this is a different area it is again of Permian strata. These agates formed in Permian strata (248-290 million years ago) that now make up part of the Krusne Hury mountains that formed during the Variscide mountain building episode. Many of these agates are more unusual in being parts of pieces of massive quartz. This occurrence is not unique to the Horny-Halze area, but has been observed only in a few other areas of the world. Nova Paka in the northwest of the Czech Republic has also been an important agate producing area for many years.

Poland

Agates from Poland have generally been available to world collectors since about 1990. Silesia in southwest Poland is an important European source of thunder egg agates. These have been shown to contain thin and long fibrous forms of microcrystals making up the agate. Research on agates from Poland by Maria Czaja showed that microcrystals forming on the flat surfaces of level-banded agates underwent less distortion than microcrystals forming on curved surfaces of wall-banded agates.

Left Note that the agate lies inside an amethyst quartz ground mass – this arrangement is exactly the opposite of what is seen in most agates. Length approximately 12.5 cm (5 in).

Opposite Agate from Breslau, Poland.

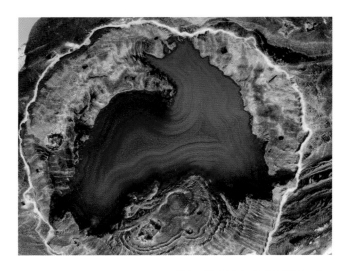

Above Chalcedony (red) inside rhyolite rock (green). This specimen is from the Esterel region of France.

France

France has not been a well-known source for agates, although some outstanding specimens have come from there. Interestingly, agate jewellery became popular in France in the years following the French Revolution. Diamonds, rubies and sapphires had been associated with the overthrown royalty, so stones of lesser value came into vogue. Agates remained popular until the Napoleonic period, when the Empress Josephine began wearing the more expensive gemstones, and brought them back into vogue. L'Esterel on the Mediterranean coast of France has been known to produce agates since about 1848, when they were first recorded by G. Coquand. These agates remained relatively unknown for many years until several different features in these thunder eggs, including a rhyolite shell, Uruguay band structure and wall-band structure, haematite bands, and a central, quartz-lined cavity, were released in the scientific press.

Britain

Britain, especially Scotland, has an abundant source of agates. The Scottish agates have been popular for use in jewellery for several hundreds of years, and they were used in some elaborate jewellery during the Victorian era.

Above *Agate-producing areas in Scotland.*

Scotland

Scottish agates are remarkably varied. Scotland probably has more known locations and varieties of agate concentrated in a relatively small area than anywhere else in Europe. Agates in Scotland occur in two types of rock: the andesitic lavas of the Early Devonian Old Red Sandstone (about 410 million years old) and the tholeiitic basalts of the Tertiary (around 50 million years old). The former are the more important of the two, and areas of andesites stretch from Stonehaven in the northeast to south of Ayr in the southwest, both areas marking the boundaries between the Highland and Southern Uplands Boundary Faults. Other Old Red Sandstone andesites surround the central granite core of the Cheviots, part of which

Opposite Agate from Montrose, Scotland.

Right Agate from Montrose, Scotland.

extends into England, whilst a small area of andesite lies just south of Oban. The Tertiary lavas are all on the islands off the west coast of Scotland, including Mull, Skye and all the Small Isles, and these are collectively known as the Small Isles agates. Another isolated area of agate-producing basalts is in Shetland. Although each locality often has its particular type of agate with characteristics unique to it, these change abruptly and given that many of the lava flows are shallow, each flow can produce a rather different variety of agate.

In total, there are at least 50 main agate areas recorded and many scattered localities within those areas. Before the discovery of Brazilian agates, many Scottish agates found their way into markets all over Britain and possibly abroad, although the famous German localities were producing agate in larger quantities from a very early time and it is unlikely that the local Scottish material ever seriously competed with the industrial production from Idar-Oberstein.

South of the Highland Fault line at Stonehaven on the northeast coast, agates have come from well-known places such as Cotbank quarry (now worked out), Den Finella (a den being a river gorge), Arbuthnott and St Cyrus to name but a few places. These agates are varied but the overall colours, whether wall-banded or onyx, are good blues to dark greys with some interesting browns and purples.

Just south of the town of Montrose, in Angus, is one of the world's most well-known locations. From Fullerton Den in the west to the tip of Scurdie Ness in the east and running down to Lunan Bay in the south is an area remarkably rich in agates of varying colour. Indeed Scurdy is a local name for the andesite and some of the older houses made out of this rock contain numerous agates still *in situ*. Onyx and sard onyx as well as pastel shaded but cracked agates appear to the south of the Montrose Basin, and easily accessible rocks lining the shore from the hamlet of Ferryden to the lighthouse (with its famous but rather less frequent 'Ferryden blue'), with warm brown wall-banded agates *in situ* and loose within the beach pebbles.

Below *Agate from Montrose, Scotland.*

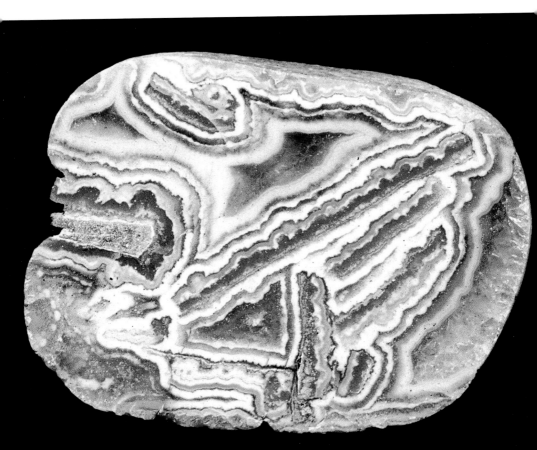

Above *Eye agate from Scurdie Ness,*
Scotland.

Around the settlement of Usan just to the south of Scurdie Ness the agates continue, particularly around 'pebble rock', the old hermit's house (whose lapidary business in the 19th century was locally famous). Rich agate country is also found in the numerous fields under plough nearby, and down the coast from Boddin to the sweep of Lunan Bay. This has been collecting ground since at least the 17th century, probably before and locals augmented their meagre wages by collecting and selling agates to Edinburgh and elsewhere, as noted by Pennant in 1769 where 'agates of a very beautiful kind are gathered in great quantities beneath the cliffs and are sent to the lapidaries in London'. Although it is true to say that there is more in the ground than ever came out of it, as far as the Usan area is concerned, most of the agate is well hidden within very hard lavas. Near Usan, situated within what is now a ploughed field, was the almost mythical 'Blue Hole', opened and worked by friends of M.F. Heddle, Scotland's greatest collector and cataloguer of minerals. It used to be called 'Mr Milne's old quarry' and produced some remarkable agates, both wall-banded and onyx. It is rare to find truly blue agates, but here they varied from royal blue to cornflower blue as well as shades of pink and brown.

Above *A Blue Hole landscape or scenic onyx agate. These agates are characteristically inky-blue and white, but some yellow and red-brown ones are not uncommon.*

Opposite *Fortification agate from Ardownie quarry, Scotland.*

Near Dundee and Moniefieth are two quarries: Ardownie and Ethiebeaton. The former has undoubtedly produced the best agate found in Scotland, some up to 60 cm (24 in) in length together with deep purple amethyst. The agates collected here, saved from the crusher, are all of a wall-banded variety of many different colours and coming from an exceptionally thick flow of andesite nearly 50 m (164 ft) in depth. The more colourful reds and oranges came predictably from the topmost section of weathered rock, whilst the splendid deep blues came from well down the flow. This was the only amygdaloidal section accessible in the quarry due

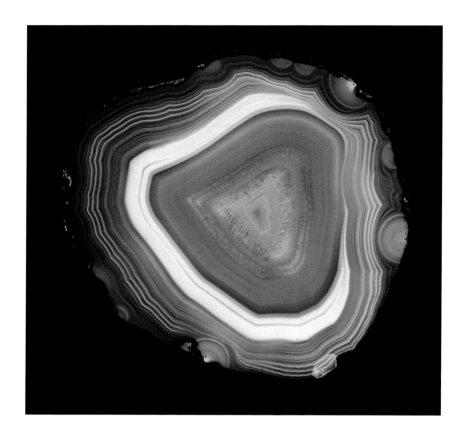

Above *This Kinnoull Hill agate from Scotland clearly shows the skin, the chalcedony layer, reddish hemi agates, grey-white chalcedony layers and a pinkish centre composed of alternating layers of chalcedony and quartz.*

to the dip of the rock, and has now been worked out. Nearby Ethiebeaton quarry produced an equally fabulous but smaller suite of agates that typically were uncracked, wall-banded, deep blues and purples with some areas of reds and browns; this too is now worked out.

Going east towards the city of Perth, traditional collecting sites in the Sidlaw Hills such as the Agate Knowe and Pittroddie Den remain as good today as they always have, being largely under plough in winter. Other less popular spots include places like Black Hill and other remote grassy hills with little visible outcrop, whilst Kinnoull

Hill near Perth used to supply much pink and orange agate in the past and is still productive in sporadic areas today, although extremely hard to work and precarious to get to.

Across the Firth of Tay and into the county of Fife lie the Ochil Hills, stretching from above Stirling in the west to just short of St Andrews in the east. They are most productive within the eastern half of the range. South of Perth the range includes rather deserted and isolated localities, famous in Heddle's time, such as the tiny settlements of Path of Condie and Path Struie. Agate is still plentiful today within the ploughed fields of the area and also within the rock in places. Ochil agates tend to be rich browns and oranges, often pinks and wall-banded with 'eyes' and a fortification pattern. Other types are white-banded greys, often of a large size, sometimes containing good quartz or amethystine geodes. All the agates are wall-banded types and onyx banding is unknown here.

Agates continue to crop up eastwards but the best known areas are situated in the North Fife Hills, an extension of the Ochils. Spread around Norman's Law, the highest of these hills, are numerous farms, some of which have been collecting

Left and below Two fortification agates from Norman's Law – a prominent hill in Fife Scotland. Few agates have been found on the Law itself, but some very fine blue-grey-white agates have been discovered, probably originating from the old quarries in the vicinity.

grounds for over a century. Every type of wall-banded agate has been found here, some of considerable size and mostly of many fine colours, varying from greys, blues through to pinks, browns and oranges distinctly patterned by thick white cachalong bands in the best localities. Onyx and sard onyx are also found but nearer the town of Cupar, where excellent yellow and red varieties have been keenly sought after. Examination of these agates show that they have been glacially transported, possibly from weathered out sources from the North Fife Hills, as the underlying rock around Cupar is sandstone.

To the southwest of Glasgow lie smaller areas of andesite, in particular near the town of Galston and within the Carrick Fell behind the coastal town of Dunure. The first of these localities is possibly unique and was the only commercially mined area for agate in Scotland. The small stream of Burn Anne runs through rich grasslands in the gentle hills behind Galston and at a few places reveals fantastically coloured vein agate of brilliant reds and yellows, mauves and blues. These brick-shaped masses of jasp agate come from long stringers – or elongated thin veins – that lie buried below the pasture, set within a matrix of green flaky clays and boles. The stringers can be long and lens-shaped with quartz at the thickest section and agate either side, the best being adjacent to the quartz and the least attractive (a dull red) at the thinnest farther ends. Pieces of these veins still occasionally erode out of the beds of the burn, but today there is no sign of the previous workings, which were at their peak in the late 18th and early 19th centuries.

Brown Carrick Fell behind Dunure, south of Ayr is rich in pink and red agates, onyx and sard onyx, as are the pebble beaches and shore rocks; a good storm is usually needed to bring fresh agate onshore. However, agates can be collected in abundance from many of the ploughed fields as well as from stream beds, shore rocks and other localities; the problem as always being not the shortage of agate but the lack of exposed outcrop to collect it from. Some of the agates from the fields are superbly coloured jasp agates that differ markedly from those found nearby in rock, and leads to an assumption that at least some of the iron oxide colouring is comparatively recent weathering of the subsurface rocks immediately below the current soils.

Agates are also found in the andesites surrounding the central granite core of the Cheviot, a large, rounded sub-range of hills partially in Scotland and England. Cheviot agates are often strongly coloured in reds and pinks with less of the ubiquitous greys. An unusual but splendid agate type of alternating red-banded agate and crystalline quartz occurs here in some of the localities, particularly on the

Above *An onyx agate from Middlefield farm in Scotland. Small numbers of agates of brilliant yellow and red colour have been found in this locality. They are quite different from most Scottish agates and highly sought-after.*

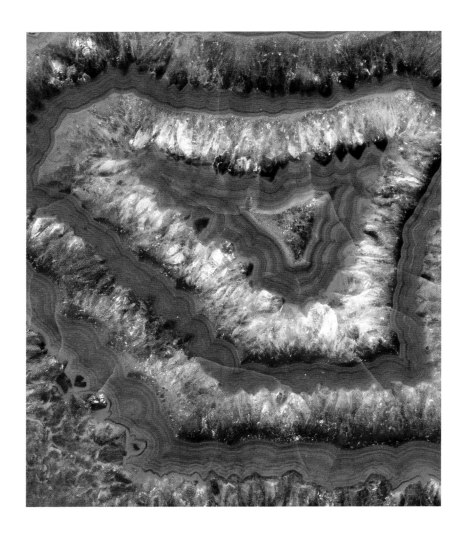

English side near the River Coquet, but also in ploughed fields on the Scottish side. Vein agates can be found in pebble banks in rivers and these can be a fine red agate surrounded by green moss. These 'border' agates are less common than those found in the Central Valley, there is less of a collecting history and new localities are quite possible given time and luck. Many Central Valley agates have been studied recently in attempts to clarify their origins, and researchers have used the ($^{18}O/^{16}O$) ratio in agates to determine the temperature under which they formed, concluding that they originated from low-temperature silica gels (see p. 12).

Collecting in the Tertiary lavas is far more adventurous, less well-known and provides a real chance of finding some entirely new locality. One of the best places is the Island of Mull, and in particular the uninhabited area on the south coast of the Ross of Mull, where cliffs of tholeiitic basalts rising to 300 m (984 ft) stretch from Shiaba to Carsaig. Agates are found in certain lava flows, sometimes indicated by the presence of zeolites either side of the agate-producing sections of flow, and at other times by the brown coloured lavas, sandwiched between black columnar basalts. The agates are often uncracked, being relatively young and usually of a deep grey/blue with thin white bands. Many have yellow calcite inclusions and most are wall-banded, sometimes with strong patterns; areas of onyx-banding can also be found. Little or no warm colours are present, although the existence of some reddish agates on nearby Iona would suggest that ferric iron-stained agates do occur

Opposite A superb but rare type of agate from the Cheviot Hills in Northumberland on the Scottish-English border. It contains concentric bands of brilliant jasp agate alternating with bands of crystalline quartz.

Below A whorl agate from Mull, Scotland, depicting fan-shaped quartz crystal groups alternating with curved bands of white or grey-white chalcedony.

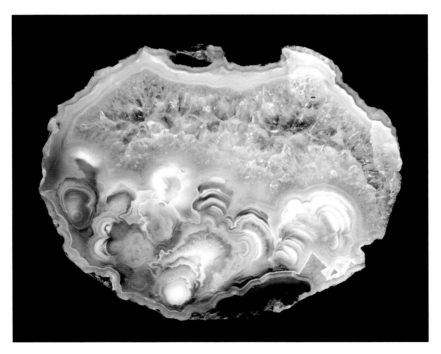

rarely. In other areas superb glass-clear rock crystal geodes are found but invariably the best of these have thin walls and are located in the most solid basalt. Agates of a similar type occur in isolated batches within Skye and Rhum (where fine bloodstones exist) in a type of lava called mugearite.

The silica supply for the Tertiary basalts is not easy to ascertain, as there is no obvious horizon of volcanic ash, save for initial deposits erupted as a first stage before the huge eruption and build-up of the main series of plateau lavas. In Skye these early horizons of ash have left deposits up to 30 m (98 ft) thick, varying from fine grained to blocks 20 cm (8 in) across. Pillow lavas, indicating deposition of lava underwater, are found at this level, and it is presumed there were once shallow lakes here. Since Skye had a Mediterranean type climate at the time, it is possible that these lakes may have contained alkaline waters that ultimately provided a silica source for agates below the current sea level and for a limited vertical distance above.

There are also rafts of sediments where possible decomposition of vegetation supplied the missing silica, particularly in Mull where most of the flows are thin and succeeded each other in quick succession with periods of quiescence where sedimentary deposits could build between eruptions in certain areas. Such an example is where the fossilized 'MacCullochs tree' lies, carbonised by a lava flow and indicative of the existence of sediments. Perhaps not entirely unexpectedly, agates occur not far from the famous tree, but they also occur in areas where no immediately obvious rafts occur either. Most of the agate horizons are within the range of sea level to 60 m (196 ft), usually with a massive assemblage of non-agate-bearing but zeolite-rich lavas above. The Tertiary agates are largely greyish-white being free from colour pigments. These agates often contain brown calcite surrounded by chalcedony and crystalline quartz.

England

'Pot stones' or 'potato stones' are irregular or sub-spherical nodules (or geodes) composed of a kind of agate that has been found in dolomitic conglomerate and the Kemper Marls of Triassic age (206-248 million years old), in about a dozen localities south of Bristol in Somerset. These stones have also been referred to as 'Dulcote agate', after the settlement of Dulcote nearby. These stones usually have an outer layer of banded agate and very fine quartz crystals which surround a hollow cavity with large quartz crystals and sometimes calcite and celestine crystal inclusions, but some are entirely agate. Potato stones are often large, perhaps reaching 20 cm (8 in) in diameter, and they may be spherical or irregular shaped.

Above *Agate from Bristol, England.*

Agates are also found in glacial drift deposits, especially at Carnelian Bay near Scarborough in Yorkshire and in the Midlands near and within the River Trent. Some poor quality agate is found in the lavas near Buxton in Derbyshire, but much better quality agate is found in the Cheviots (as previously mentioned), particularly around the River Coquet.

Some delicate pink agate has been found near Lynmouth, Devon, and some agate nodules with geometric figures such as pentagons and hexagons on their surfaces, which are reminiscent of thunder eggs, have been found along beaches near Clevedon, Somerset.

Above *Potato stone agate from Dulcote quarry, Somerset, England.*

The collection at the Natural History Museum, London, includes agates from localities near Liskeard and St Agnes in Cornwall, Wheal Friendship near Tavistock in Devon, the Fernley Pit near Lichfield in Staffordshire, several localities near Birmingham in Warwickshire, the beaches near Brighton in Sussex and the beaches near Dover in Kent.

Wales

The Ordovician of Wales (443-495 million years ago) has produced some very interesting thunder eggs that were first studied in detail in 1889 by a geology student named Catherine Raisin. Raisin recorded agate nodules (thunder eggs) from Pen-y-chain and Carig-y-defaid in North Wales, on the Lleyn Peninsula, and she recognised that these formed in altered volcanic ash with glassy shards of obsidian (a rapidly cooled lava). This same observation was made by other researchers in 1961 when studying the origins of thunder eggs in Oregon. Raisin also observed that the thunder egg nodules she studied had colour zonations that in 2001 were related to slight differences in mineralogy of the outer and inner shell of the nodule. Perhaps the most important feature of Raisin's work was that she was the earliest worker to relate agates to their environments of deposition. She was an astute observer and interpreter of data, and reached a high plateau in science long before most women entered scientific careers. Unfortunately for agate researchers, Raisin did not follow up on her 1889 study. Agate has also been found Llanrhaidr in North Wales.

Right Agate from Llanrhaidr in Wales.

USA

The northeast states

Early sources of agates in the USA included the states of Connecticut, Rhode Island, New York and New Jersey. Literature shows that these sources were known by the mid-1700s, and collections of these agates have been documented from as early as 1820. Unfortunately, for agate collectors, this part of the USA quickly became settled and developed. The once-productive New Jersey agate-bearing deposits are almost all entirely covered by suburban developments. A few nodules of carnelian agate are occasionally found near Sterling Brook in New Jersey, but no other extensive agate deposits can be accessed.

Many fine agates were collected in the area that is now under New York City, and the Empire State Building has its foundation in agate-bearing basalt of Triassic age (206-248 million years old). Agates from the northeast states have become so rare that an advanced collector is overjoyed to find one in a dealer's stock or in an old collection. Although research collections of some museums contain some of these agates, their display collections rarely include any of them, as they are often overshadowed by larger, more spectacular specimens from Mexico or Brazil.

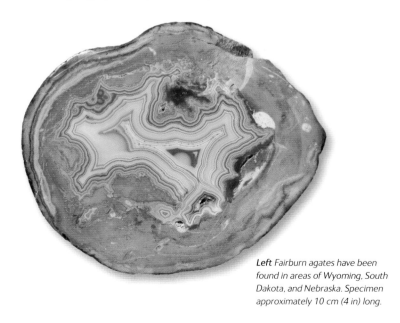

Left *Fairburn agates have been found in areas of Wyoming, South Dakota, and Nebraska. Specimen approximately 10 cm (4 in) long.*

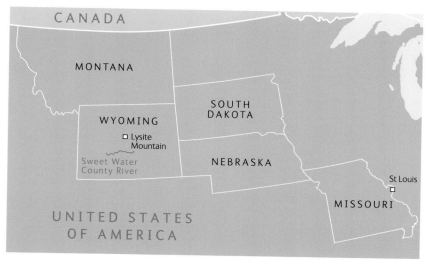

Above Map of USA midcontinent.

The midcontinent

The midcontinent has been the source of many agates, most of which have been found hundreds of kilometres from their sources, in stream or glacially deposited sand and gravel. Such agates have been very popular with collectors inasmuch as the agates are concentrated in one area and perhaps close to the surface.

Late Carboniferous and Permian rocks (248-330 million years old) exposed in Montana, Wyoming and South Dakota have yielded some outstanding agate specimens. The older of these agates that formed in the Late Carboniferous limestone of Wyoming and South Dakota are called Guernsey Lake agates in the former area and Tepee Canyon agates in the latter, although they are generally parts of the same population of agates. These agates were first recognised in gravel derived from the agate-bearing deposits, and were laid down in parts of South Dakota and Nebraska, in the Oligocene epoch of the Tertiary (24-37 million years ago), where they are commonly called Fairburn agates, after the settlement of Fairburn, South Dakota. Structurally, these marine sedimentary agates are similar to amygdaloidal agates, except they lack most of the inclusions that are common to the latter. Zeolite minerals, crystal pseudomorphs, plumes and other such inclusions common to amygdaloids have not been observed in these marine sedimentary agates.

Above Union Road agates have been found in
Lower Carboniferous (Mississippian) limestone in
eastern Missouri. Many of them show a fibrous
structure on the outer bands that grades into a
granular structure toward the interior of the nodule.
Specimen approximately 7.5 cm (3 in) in diameter.

Another major occurrence of marine sedimentary agates in the midcontinent is in eastern Missouri, near the city of St Louis. These agates have all been found in place in limestone of Early Carboniferous age (330-354 million years old). These agates are called Union Road agates. Many of these agates are characterised by being incompletely banded, the banded portion of the agate being situated toward the outside of the nodule. The banded part of the agate more or less imperceptibly grades into granular chert.

Continental sedimentary agates have been found in place in parts of Wyoming, South Dakota and Nebraska. These kinds of agates formed in cavities, probably created by burrowing animals, and fissures formed by faults and joints in wind-deposited siltstone and claystone, and it has been suggested that the silica source for these agates is volcanic ash.

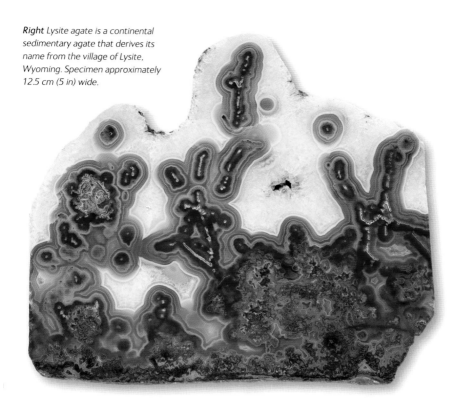

Right Lysite agate is a continental sedimentary agate that derives its name from the village of Lysite, Wyoming. Specimen approximately 12.5 cm (5 in) wide.

Lysite agate derives its name from the village of Lysite or Lysite Mountain, Fremont County, Wyoming, and it is commonly a very colourful agate. In addition to bands, Lysite agate may contain plumes and moss that produce very attractive specimens. To date, all of the specimens of Lysite agate that have been observed have been vein agates and no nodules have been observed.

Sweetwater agates have derived their name from either Sweetwater County or Sweetwater River, Wyoming, where most specimens have been collected from sand and gravel deposits derived from nearby porous sandstone of Miocene age (5-24 million years old). These are generally small moss agates that contain dark brown to black dendrites of oxides of manganese and/or iron. Most of these agates fluoresce under long wave ultraviolet light, and many collectors prospected for these stones at night. Sweetwater agates are usually very well rounded when found, and are tumble polished and made into inexpensive pieces of costume jewellery. These agates were being used as early as the late 1920s, where they were commonly sold at tourist stops on the way to Yellowstone and Grand Teton National Parks.

Right Sweetwater agates are continental sedimentary agates. They are generally granular and may contain many dendritic patterns. They are highly fluorescent under longwave ultraviolet light, and contain more uranium than other agates. Largest specimen is approximately 2.5 cm (1 in) in diameter.

Nebraska blue agate is an abundant form of agate that covers much of the surface in several areas in northwestern Nebraska and southwestern South Dakota. It formed in fault and joint planes in wind-deposited siltstone and claystone of the Chadron Formation of Oligocene age (24-37 million years old). Although this agate is commonly called blue agate, most of it is grey in thinner slices (of 6 mm or ¼ in). Thick pieces may show an attractive blue colour, but these shades quickly fade when the specimen is cut into thinner slices. These agates have many interesting inclusions and structures that are suitable for making doublet gems (thin slices of agate covered with a transparent quartz or glass cap cut *en cabochon*). Only a very small percentage of the agates found in this area is suitable for either cabochons or specimens, either because the pieces are too small, too highly fractured or too porous.

Above *Historically, the term chalcedony has been used to describe light grey to colourless, banded, cryptocrystalline quartz. This specimen of Nebraska blue-agate is approximately 10 x 12.5 cm (4 x 5 in).*

Right *Moss agates that formed in continental sedimentary deposits of the Harrison Formation of Miocene age and the Ogallala Formation of Pliocene age have been found in the high plains of the USA for over 100 years. Specimen approximately 7.5 cm (3 in) long.*

A kind of agate from near the same area where blue agate has been collected and from adjacent areas in Wyoming is a moss agate that is found near the top of the Miocene Harrison Formation (24-26 million years old). Much of this material shows a gradation from common opal with dendrites to chalcedony or agate with dendrites. These agates were sometimes used by Native Americans to make stone tools and were commercially mined for jewellery in the late 1890s and early 1900s.

North central states

Agates have been found both in basaltic rocks of Late Precambrian age (over 545 million years old), and in glacial and stream deposits in parts of Minnesota, Wisconsin, Iowa, Missouri, Nebraska and Kansas in the USA and in Ontario, Canada. These agates have been eroded from their host rock and have been transported and re-deposited in glacial deposits of Pleistocene age and in Recent stream deposits. Copper mining logs, that have been made by both private mining companies and by government geological surveys, show that there are numerous different layers of basalt or trap rocks that have yielded agates in place. Lake Superior agates are a 'family or group of agates' of multiple origin in host rocks that range in age from about 1200 million years to 600 million years old. They may include some of the world's oldest agates. So far, only thunder egg and amygdaloidal agates have been observed in the Lake Superior agates. These same agates have been found along the Mississippi River as far south as Arkansas and Louisiana.

Opposite *Agate-producing areas in north-central USA.*

Above *Sagenitic inclusions in Lake Superior agates may include needle like crystals of zeolite minerals, alteration products of basaltic rocks. Specimen is approximately 6 cm (2¹/₂ in) long.*

Below *Example of an unweathered Lake*
Superior agate from a mine in the Keweenaw
Peninsula of Michigan.

Such agates have commonly been called Lake Superior agates, a name derived from the Lake Superior Till, a Pleistocene glacial deposit (1-1.8 million years old) in Minnesota, rather than from Lake Superior itself. In the late 1800s and early 1900s, these agates were frequently picked up along the beaches of several of the Great Lakes. Many were made into costume jewellery and an agate marble industry, based on the once-popular outdoor game, flourished in the area for several decades.

Lake Superior agates have a long and complex history. Because many of them have been transported by streams or glaciers to areas far from their host rock, they have lost many of the features that are common to other kinds of agate. Host rock is rarely attached to any of these agates, and structures or inclusions such as sagenitic inclusions and plumes are rarely preserved, as these are porous, soft, and easily broken apart by weathering and transport. Some of these agates have been found in place in the pre-Cambrian basalt or in the Pleistocene till in such areas as the Keweenaw Peninsula of Michigan, Isle Royale National Park, Michigan and Michipicoten Island, Ontario, Canada. The unweathered agates that have been found in place are commonly opaque, whereas the agates that have been removed from their host rock and subjected to weathering and transport are normally translucent.

Southeast states

The southeast states have provided some exceptional agates, most of which are of marine sedimentary origin. The states of Tennessee and Alabama have produced some very attractive agates. Florida is well-known for its agatised coral from Tampa Bay.

Several kinds of agates have been documented from Tennessee and Alabama. As well as carnelian and iris agates, those such as the Calf Killer agate, Horse Mountain agate and Paint Rock agate, are named after local geographic locations (though no Calf Killer River is listed by the USGS). Those from Horse Mountain in Bedford County, Tennessee, are of Ordovician age (443-495 million years old). The agates from the Alabama sources are very colourful and often contain some plume-like inclusions.

Above Paint Rock agates are usually very brilliant shades of red and yellow, and some may get quite large. Diameter of specimen approximately 8.7cm (3^1/$_2$ in).

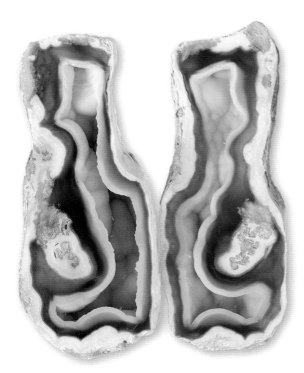

Left *Agatised corals from Tampa Bay, Florida, are mostly shades of yellow, but some specimens have black bands with a botryoidal (shaped like a bunch of grapes) structure inside. Height approximately 7.5 cm (3 in).*

Below *Slices of hollow Tampa Bay coral are very popular for jewellery objects. Diameter of largest slice approximately 2.5 cm (1 inch).*

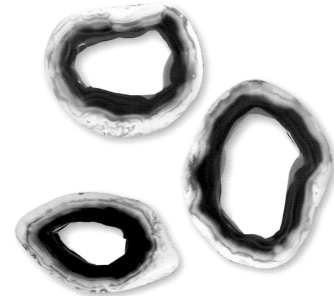

Tampa Bay in Florida has produced some exceptional examples of agatised coral. These have been known since the early 1800s, and may have been used as gems as early as 1825. Because of environmental restrictions, the classic localities for these corals right in Tampa Bay itself have been inaccessible to collecting since about 1975. Many examples of these corals come to light as the result of excavations for new construction on the land, and private companies contract with the builders for the rights to excavate the coral producing horizons. Strom, Upchurch and Rozenweig (1981) suggested the agatisation of these corals resulted from silicification at the surface. In addition to the coral fossils themselves, Tampa Bay agatised corals may contain the remains of other marine invertebrates such as clams and sand dollars (echinoids). Organic silica derived from sponge spicules and radiolarians that were contained in the marine sediments, provided the source of silica when the coral bearing rocks became exposed to the air and deeply weathered. Similar agatised corals have been found as far north as the Suwannee River in Georgia and northern Florida.

The Great Basin and southwest states

California has yielded a large variety of agate, and in the early part of the 20th century many prospectors had been successful at finding deposits. California has produced mostly vein agates, which are agates formed in large fractures in bedded

Below *Agate-producing areas in the Great Basin and southwestern USA.*

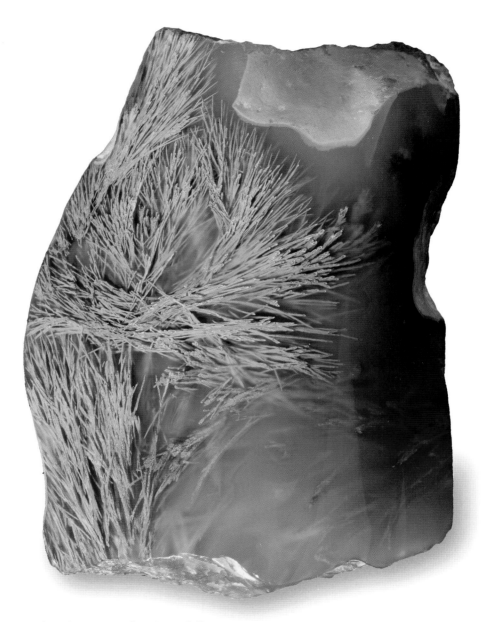

Above This specimen is from Nipomo, California.
Sagenitic agates have also been called bean field agates
inasmuch as they were first collected from freshly
ploughed bean fields.

rock rather than in vesicular cavities in the rock. Nodular agates are not unknown in California, but they are less common than the vein agates or 'veined silicates' as some are popularly known.

Nipomo or bean field agates have been found close to the Pacific coast, near Nipomo, north and west of Los Angeles. The agates were collected from bean fields, and the supply proved almost inexhaustible as long as beans were the principal crop of the area and tillage exposed fresh material annually. Most of the area is now planted in fruit orchards, and Nipomo or bean field agates have all but disappeared from shows and dealers' stocks. These were generally small, sagenitic agates with yellow or brown needle-like inclusions. Many of the historic localities in southern California have been lost to either changes in crops or urban growth and sprawl.

Many of the agates found in the Great Basin area of California have been named after the localities from which they were collected. Most of the agates are vein agates, but just about every variety of agate has been collected. Lavic Siding is a railroad siding that lends its name to Lavic jasper and Lavic agate. This is one of southern California's prime localities. Lavic agates commonly have brilliant red plumes. Horse Canyon agates are commonly moss agates or plume agates that have been found in Tehachapi Canyon. The canyon derived the nickname Horse Canyon because some of world's oldest fossil horses have been found there. Paisley agate has been collected in several areas in the Great Basin and in Mexico, and the California material may be red swirls in a white to light grey matrix, or dark green and light grey on a brown matrix. Paul Bunyon agate seems to be somewhat of a misnomer since Paul Bunyan was a fictional to near mythic figure in the history of the forests of the Great Lakes areas along the USA-Canada border. Siam agate is

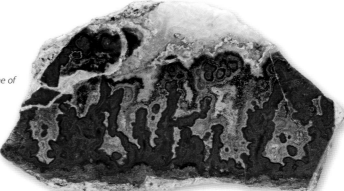

Right The origin of the name of this colourful Paul Bunyon plume agate is not known, and the spelling is different from the name of the North American folklore character Paul Bunyan. Specimen is approximately 10 cm (4 in) long.

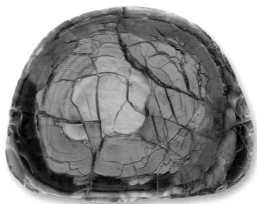

Above *Siam Siding in California is another railway stop that provided a name for agates and other quartz family gems. This material is sometimes called Chapinite – an invalid mineral name. Specimen is approximately 8 x 10 cm (3¹/₂ x 4 in).*

Left *Wingate Pass is near Death Valley in southern California. This site has produced fine plume agates for many years. Specimen is approximately 10 x 11 cm (4 x 4¹/₂ cm).*

Below *China Lake in California is no longer accessible to collecting, but it has yielded some outstanding plume agates in the past. Slice approximately 20 cm (8 in) long.*

Right *Fourth of July Butte gives its name to several kinds of banded agates. Slice approximately 10.2 cm (4 in) long.*

a brecciated agate that has been found near Siam Siding, another famous locality in California. Wingate Pass near Death Valley has produced some outstanding agates that have red and yellow plumes; these are sometimes called Death Valley agate. Owl Hole Lake is known for fine sagenitic agates.

Southern California has had many outstanding agate-collecting localities. Unfortunately, most were lost to the expansions of the boundaries of military posts during the Second World War. China Lake, Bicycle Lake and Lead Pipe Springs are several of these Californian localities that are now unavailable to collectors.

Arizona has several important agate-producing localities that have produced some outstanding agates, but these have received somewhat less attention than the California localities. Fourth of July Butte has offered up some fine, commonly grey, banded agates, and the area near Clifton has produced some outstanding translucent to opaque, purple and white agates. The area around Burro Creek is known for sagenitic, dendritic and moss agates. The area around Brenda, Arizona, has also produced many agates.

Nevada has produced its share of agate, and material from two localities is currently popular. Bull Canyon is in the northwestern part of the state and is well-known for plume and moss agates that come from veins. Amethyst sage agate is a dendritic agate that is in a violet to yellowish matrix and is very popular among jewellery makers. Some beautiful black and white thunder eggs have been found in the Black Rock Desert in northwestern Nevada. Groom Lake is reputed to have produced agates, although access is not granted as it is a military facility.

Utah is best known to agate collectors for a kind of thunder egg called a Dugway geode. These stones are commonly hollow and have a lining of very fine quartz crystals that are within a light grey to blue-grey, banded agate shell. Some specimens are completely filled with agate, but these are relatively rare. Utah is also known for Pigeon Blood agate, an agate that is characterised by having blood-red, mossy inclusions in a relatively clear and colourless matrix.

Right Bull Canyon in Nevada is the source of some very fine, colourful moss agates. Specimen is approximately 12.5 x 15 cm (5 x 7 in).

Left Black Rock Desert in northwestern Nevada is the source of some very beautiful black and white agates that are still in their volcanic matrix. Specimen is approximately 17 cm (7 in) wide.

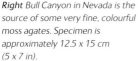

Below The area around Deming, New Mexico, is a popular agate-producing area in southwestern USA. There is even a state park where collectors are allowed to dig for agates. Specimen approximately 6.3 cm (2¹/₂ in) in diameter.

New Mexico has a number of important agate-producing localities, the most important of which is near Deming, a site that was still productive in 1995. The state of New Mexico has established Rockhound State Park, an area where collectors can pay a fee for the privilege of prospecting and digging for agates and other gems. Most of the agates are relatively small thunder eggs, most of which are about 5 cm (2 in) in diameter, although some large examples have been found. Some vein agates and plume agates have also been collected in this area.

West Texas has been a famous agate-producing area for about 100 years, and the first examples of these plume agates, from south of Alpine, made their appearance at the Trans-Mississippi Exposition in St Louis Missouri, in 1901. The agates are locally named after the ranches from which they were collected, such as Woodward Ranch, Walker Ranch, etc., but all appear to be part of the same group of agates, all being collected from the Cottonwood Spring Basalt, a series of lava flows that are of Miocene age (about 23 million years old). The most highly desired agates from west Texas are the plume agates that have fine, feathery inclusions that may be black, red or even a golden colour, the latter colours being the most desired. A. Zarins studied these agates in detail in 1977 and concluded that their silica source was volcanic ash and that they commonly formed in areas where ancient alkaline lakes once existed. In addition to the plume agates, moss agates, banded agates and thunder eggs that are sometimes called pompon agates have been collected in the area between Alpine and Big Bend National Park. Many of the agates that formed in west Texas have been eroded from their sources and subsequently transported towards the Gulf of Mexico by the Rio Grande River. In the eastern part of Texas, these are called Rio Grande agates.

Below Some thunder eggs have been found in the Alpine area of Texas, and they are commonly called pompon agates. They often have radiating inclusions of an as yet unidentified zeolite mineral. Specimen approximately 10 cm (4 in) long.

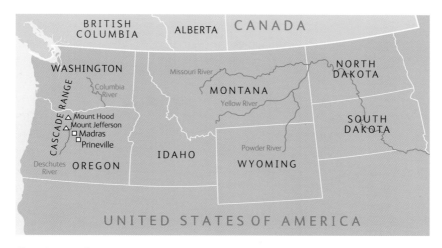

Above *Agate-producing areas in the northwestern USA.*

The Pacific northwest

This part of the USA includes Montana, Idaho, Oregon and Washington, most of the latter three states being west of the continental divide. The northernmost parts of these agate-bearing terrains extend into British Columbia and Alberta provinces in Canada.

Montana is known among agate collectors for Montana moss agate, a gem that has been known in the trade for over 100 years. Montana moss agates usually range from colourless to showing shades of translucent grey or yellow body colours that may have very exquisite patterns of dendrites, sunbursts or scenes that are composed of mineral inclusions of oxides of manganese and/or iron. Montana moss agates comprise a complex group of agates that represent a multitude of sources from many areas and several geological ages. Some of the Montana agate was probably derived from agatised limb casts – mineral fillings of voids formed when the limbs or trunks rot out of sediment or burn out in hot volcanic ash formed in volcanic rocks in west central Wyoming and parts of Idaho. Montana agates with marine invertebrate fossils of Devonian age (354-417 million years old) have also been recorded. All of the Montana agates have been found as well rounded, water-worn nodules. They have been found in gravel deposited by the Missouri River as well its tributaries such as the Yellowstone and Powder Rivers, not only in

Above Montana moss agates originated in several
sources and most are found in gravel of the
Yellowstone River and some of its tributaries. The
colouring is due to oxides of manganese and/or iron.
This example shows two different kinds of colouring.
Length of specimen approximately 8.8 cm (3¹/₂ in).

Montana, but in North and South Dakota, and as far away as southeastern Nebraska.
Some Montana agates exhibit the unusual property of appearing absolutely void
of dendritic patterns when viewed in transmitted light from one direction, but have
abundant dendrites when viewed in transmitted light from the opposite direction.
This may be a fibre-optical property, but no cause for this phenomenon has been
determined.

Idaho has many different kinds of agates. Many of the localities are near its border
with Oregon, and some sites such as Succor Creek have essentially been shared by
the two states. Graveyard Point near Nyssa has been the source of fine plume agates
for many years. Some of the agates from here that have a pinkish cast have been

Above Graveyard Point on the Idaho-Oregon border has yielded fine examples of plume agates for many years. The name Regency Rose agate is applied to these agates that have red or pink plumes. Many of the plumes contain pyrite. Specimen is approximately 8.7 cm (3¹/₂ in) high.

called Regency Rose agate. These are vein agates and some rather large pieces have been found. Sagenite Hill derives its name from the fact that sagenitic agates have been found there in great profusion. These sagenitic agates are the cores of thunder eggs, and most pieces have very irregular outlines. Green and white moss agate has been found in the Muldoon area, and many of these provide very beautiful specimens. Beacon Hill agates appear to be mostly amygdaloidal agates that have regular outlines. They are usually shades of light grey and white and may have many inclusions.

Oregon is covered by extensive deposits of rhyolitic volcanic ash and basalt flows, and because of this complex igneous terrain, Oregon has yielded more different

Above The area around Muldoon in central Idaho has yielded some colourless agates that have green chloritic or micaceous inclusions as well as white stalk aggregates. Specimen is approximately 6.2 cm (2^{1}/$_2$ in) long.

kinds of agates than any other place in the world. Agate occurrences were known in Oregon as early as the late 1800s, and many of these agates reached cutting centres throughout the world. A knowledgeable collector of agates who visits the jewellery counter of any antique shop will likely spot several examples of agates from Oregon in a Victorian or Edwardian setting. The surface rocks to the east of the Cascade Range in Oregon are made up of numerous beds of welded ash flow tuffs, whereas the rocks west of the Cascades contain mainly basalt flows. As a result, the agates from eastern Oregon, mostly thunder eggs, are strikingly different in character than those from western Oregon, which are mostly amygdaloidal agates. The best known agates from Oregon are probably the famous thunder eggs from the Blue Bed or Pony Butte on the Richardson Ranch (formerly Priday Ranch) northeast of Madras, Oregon. These agates are found in a decomposed and devitrified layer of grey, rhyolitic volcanic ash that is included in the John Day Formation of Miocene age (5-24 million years old). Popular legend tells the story of these thunder eggs representing missiles hurled at each other by feuding gods on Mount Hood and Mount Jefferson. This legend appears to have been created by tourist shop vendors, as it has not been located in any texts on Native American mythology. These thunder eggs have a dark brown shell that is usually filled with blue and white agate. Both wall lining and horizontal or Uruguay banding have been observed in these gems. Some of the Blue Bed thunder eggs that have been found weathering on the surface may have more attractive colours such as red or yellow inside. This is probably due to the oxidation of iron impurities within the agate structure.

Above Double chambered Blue Bed thunder egg. Note that there is no similarity between the patterns of agate in either chamber. Diameter approximately 12.5 cm (5 in).

Above Fine plumes in Priday plume
agates are commonly restricted to the outer
surface of the nodule. The back side of this
specimen is entirely matrix showing no plumes.
Diameter approximately 6.7 cm (2$^{1}/_{2}$ in).

Another kind of agate that has been found on the Richardson Ranch is the Priday plume agate. These agates have sometimes been called Fulton plume agates; this name was used when the ranch was operated by the Fulton family. The finest Priday plume agates have brilliant red and yellow plumes in a nearly transparent matrix, and these are highly prized by both agate collectors and jewellers. Fine specimens may command very high prices. The thunder eggs that contain the Priday plume agate are similar in outward appearance to the Blue Bed thunder eggs. The plumes may be a very localised occurrence in a bed that correlates with the Blue Bed. The Priday plume bed contains numerous agate nodules, but only a small percentage of them actually have fine plumes.

The Frieda thunder egg bed is situated nearer to Prineville, Oregon, and it produces thunder eggs that commonly have green, chlorite or magnesian chamosite inclusions in them. This locality is one that has been important since as early as the 1920s, and many old collections have specimens of Frieda thunder eggs in them. Eagle Rock plume agate derives its name from a topographic feature of the same name near Prineville in Crook County, Oregon. These agates have been offered as both Eagle Rock plume agate and Angel Wing agate. The name Angel Wing has become somewhat generic, and applied to many plume agates regardless of origin. They are usually vein agates with red, pink or black plumes in a relatively translucent matrix. Cabochons cut from Eagle Rock plume agates are often made into doublets, thin sheets of the translucent stone cemented to a clear quartz cap to allow more light to pass through the agate to better display the plumes.

The Antelope area in Oregon contains several interesting varieties of agate. The agates have been offered under the names of Antelope agates and Big Muddy Ranch agates. There are some differences in the inclusions observed in these agates. Many of the stones that have been offered as Antelope agates are pink and have membranous cristobalite inclusions. Stones that have been offered as Big Muddy Ranch (also Muddy Ranch) agates often have stalk aggregates filled with a green, clayey mineral that is possibly magnesian chamosite.

Deschutes River thunder eggs have been found in beds along the stream that leads from its source in the Cascade Range in southwestern Oregon to its confluence with the Columbia River in north central Oregon. These thunder eggs are commonly large, 15 cm (6 in) or more in diameter, and black in a clear, colourless matrix. They have the dark brown outer shell as do most other thunder eggs.

Buchanan thunder eggs derive their name from the small settlement in Harney County in southeastern Oregon. Buchanan thunder eggs commonly have a greenish

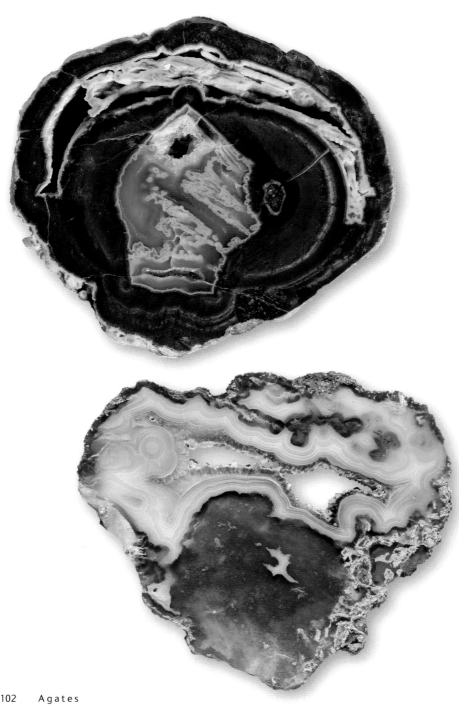

Opposite top Double chambered Frieda thunder egg. Note that the lower chamber is stelliform in outline and the upper chamber is sickle-shaped. Diameter approximately 7.5 cm (3 in).

Opposite bottom Antelope agates commonly have membranous cristobalite inclusions. Diameter approximately 12.5 cm (5 in).

matrix and usually do not have a sharply defined agate pattern in their interior. Although they are thunder eggs, they have a somewhat different outward as well as inward appearance from other thunder eggs from Oregon. Most have examples of Uruguay banding and wall lining bands are not common in these agates. They have a diameter of about 10 cm (4 in). Succor Creek is situated along the Oregon-Idaho border and its course flows through both of these states (the name is often misspelled as 'Sucker Creek'). This has been a popular collecting area for at least 70 years. In addition to agate-filled thunder eggs, the area around Succor Creek has also produced outstanding jasper with brilliant colours and well-defined scenes.

Rattle snake eggs is the name given to a peculiar kind of small, white to light grey, non-banded, granular, nodular chalcedony that has been found in volcanic rocks in southeastern Oregon. That these nondescript stones may have come into high demand is based on the fact that the agate nodule is of uniform porosity throughout. These kinds of agates have been cut into slices, and by a patented process, zinc dendrites have been introduced into the agate to produce laboratory made dendrites, or moss agates. These stones have been sold on the market as Fischer stone®.

Trent sagenite agates derive their name from the village of Trent, Lane County, Oregon, and were first collected from a railway cut near there. These agates have inclusions of black, metallic stibnite (antimony sulphide) crystals that may appear blue on very thin edges.

Holley blue agate is an amygdaloidal agate that has been found near the village of Holley in Linn County in west-central Oregon. Most are really more a lavender colour than blue, and this reflects the desire of agate collectors to have the rare blue agate. They are commonly small, usually around 2.5 cm (1 in) in diameter, and rarely reach 5 cm (2 in) in diameter.

Above *Succor Creek thunder eggs are often characterised by having very irregular agate patterns inside a dark brown matrix. Almost all are blue or light grey on the interior. Diameter approximately 7.5 cm (3 in).*

Oregon Beach agates are one of Oregon's most popular gems. Similar agates have been found along the Pacific coast from northern California to Washington, British Columbia in Canada and Alaska, but they are more heavily concentrated in the vicinity of Yachats and Newport, where they are commonly referred to as Yachats Beach agates or Newport Beach agates. These are mostly amygdaloidal agates that have been weathered from basalt flows.

Copco dendritic agates are small, amygdaloidal agates that have been found in southwestern Oregon and neighbouring northern California. The name Copco

Above *Oregon Beach agates are commonly small, less than 3.7 cm (1¹/₂ in) in diameter, although some outstanding specimens may reach twice that size. Largest specimen approximately 3.1 cm (1¹/₄ in) in diameter.*

comes from California, Oregon and the Pacific Cattle Company that operated many of the ranches in that area. These agates may have exquisite tree-like patterns of black manganese or iron oxide.

Sweet Home, Oregon, is well known for some excellent examples of black and white fortification agates that derive their name from the city of Sweet Home in Linn County. Some of these agates become as large as 15 cm (6 in) in diameter. They may contain inclusions of membranous cristobalite or other minerals such as magnesian chamosite.

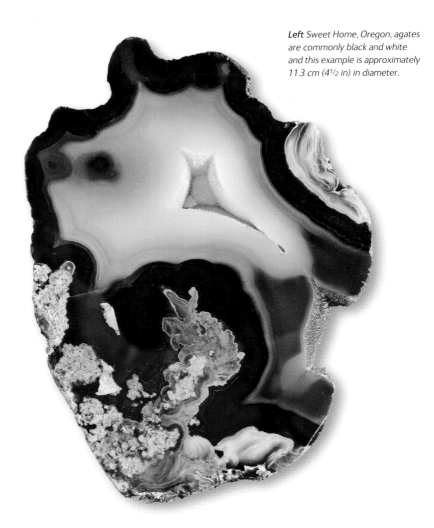

Left *Sweet Home, Oregon, agates are commonly black and white and this example is approximately 11.3 cm (4½ in) in diameter.*

Above *Sagenitic agates from near Brownville, Oregon, often have yellowish needle-like crystals that are probably of a zeolite mineral. The sagenitic inclusions in the Brownville agates show better in thin section, whereas the sagenitic inclusions in the Thistle Creek agates show up better in thick sections.*

Thistle Creek is immediately northeast of Sweet Home, Oregon, and it lends its name to some very fine sagenitic agates. The sagenitic inclusions are usually white, and contrast sharply with the dark grey matrix of the agate when they are cut into thick slices or half nodules. Brownville, Oregon, is situated in the western point of a rich agate-bearing area defined by the triangle formed by the settlements of Lebanon and Sweet Home to the east, and Brownville to the west. Some outstanding examples of sagenitic agates have been found there.

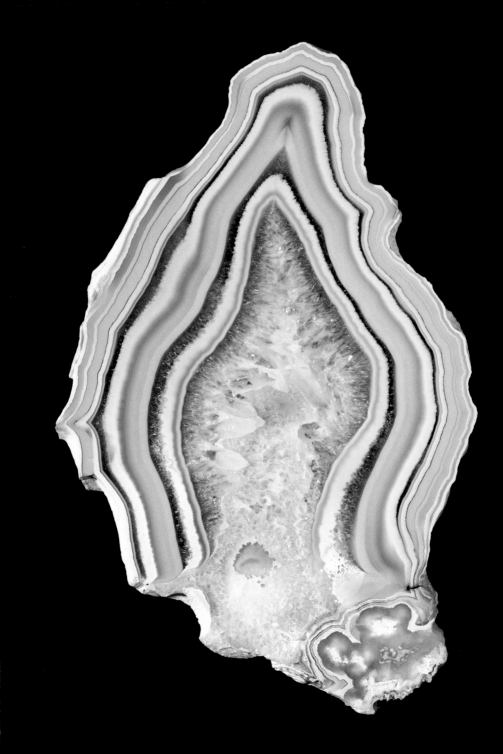

Canada

Nova Scotia is a very well known agate-producing area and it has produced an abundance of stones for well over 100 years. These agates include amygdaloids, vein agates, some thunder eggs, banded agates, moss agates and sagenitic agates. Most of the agate-producing areas are beaches and cliffs along the Bay of Fundy; there the tidal bore is normally 3.5 m (11 ft), but can reach 16 m (53 ft). This keeps the deposits in a constant flux and quickly erodes new material from the source rocks. The agates formed in basaltic rocks of Triassic age (206-248 million years old) along a rift that extends from Nova Scotia as far south as New Jersey in the USA. Agates have been found along most of this rift, but large metropolitan areas in the USA have covered most of the localities there, leaving Nova Scotia as one of the few sources for these gems. A well-known local Bay of Fundy agate is a unique variety called straw agate. Straw agate is made up of numerous, somewhat randomly

Opposite Nova Scotia has been an important agate producing area, and many localities have been recorded along the Bay of Fundy. The area has produced some outstanding agates, many of which were described as early as 1870. Specimen approximately 10 cm (4 in) high.

Below Agate-producing areas in Canada.

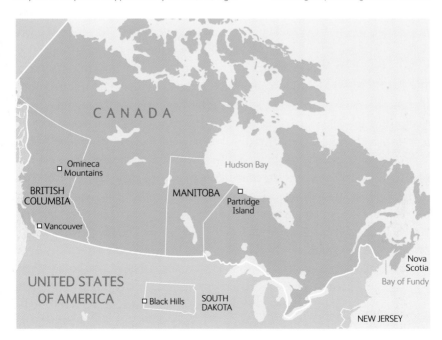

Right Nova Scotia's agates come in many varieties and may contain inclusions such as green chlorite or mica minerals, or pseudomorphs of tabular crystals of an undetermined mineral. White cabochon approximately 5 cm (2 in) long.

oriented crystals of probably a zeolite mineral. Most straw agates are grey to black, but specimens of yellow that resemble natural straw are highly desirable. An agate similar to straw agate has been found along the Colorado River in Arizona; however, there it is called lattice agate.

Manitoba, located in the central part of Canada, and bordered by the USA on the south is very well known among rock and agate collectors for a kind of agate called Souris agate, after the city of Souris near where these agates have been found. Souris agates have a very complex geology, and they come from mixed sources. The area has been heavily glaciated, and this accounts for agates from northern and northwestern sources, as well as possibly some material from the northeast, near Hudson Bay. In addition to the agates derived from the above sources, some of the agates found here have come from the Rocky Mountains where they were eroded from the source rock and transported by eastward flowing streams. Some workers have suggested that some of the Souris agates may have come from the Black Hills to the south in South Dakota and Wyoming. Many of the Souris agates strongly resemble the famous Montana moss agates that have been found in the Missouri River drainage and its tributaries such as the Yellowstone and Powder Rivers. Both of these kinds of agates may have been derived from the same sources.

British Columbia contains numerous agate varieties that include both thunder eggs and amygdaloids. The area around Omineca has yielded some outstanding agates, but little of this material seems to reach other parts of the world. Some interesting thunder eggs have also come from here, and again, few of these get to

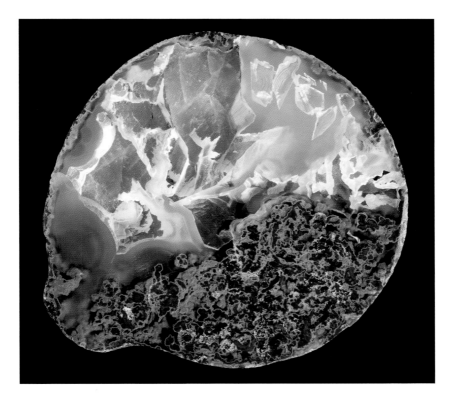

Above Omineca agates derive their name from the town of Omineca or the Omineca Mountains near Francois Lake in British Columbia, Canada. Specimen is approximately 4 cm (1¹/₂ in) in diameter.

other places. Dallasite is the name given to a green and white silicified pillow basalt, and it derived its name from Dallas Avenue in the City of Vancouver where it was first discovered.

Mexico

Because of their beautiful and unique colours, intricate structures, contrasts, and relative freedom of flaws, the agates of northern Mexico are among the most desired of all agates in the world. Agates from here were first recorded by an anonymous author in 1913, and although a few agates reached the USA before the Second World War, they did not become well known until about 1950. Now, many

Above *Agate-producing areas in Mexico.*

of the finest specimens go to Europe where they have graced both private and museum collections. These agates have been well described and type areas have been designated for many of the major agate varieties; information is also available on their mining and geological history. Although fine agates are mostly reported from the state of Chihuahua, many excellent specimens have also been collected in the states of Durango and Sonora. Most of these agates have been found in volcanic rocks that have been mapped as Tertiary (Miocene and Pliocene ages, 1.8-24 million years old) but at least one kind, lace agate, has been found in sedimentary rocks of Cretaceous age (65-142 million years old).

In the early years of post-war production, essentially all of the agates from northern Mexico were lumped together and simply called Mexican agates. The mining and sales were carried out almost exclusively by American dealers in league with Mexican companies. There are many named varieties of agates from northern Mexico, and it is difficult to establish when these names came into use. Although the names of several agate-producing areas appeared in popular publications by about 1960, actual variety names appeared somewhat later, and most of them seem to have appeared first in advertisements in gem trade journals.

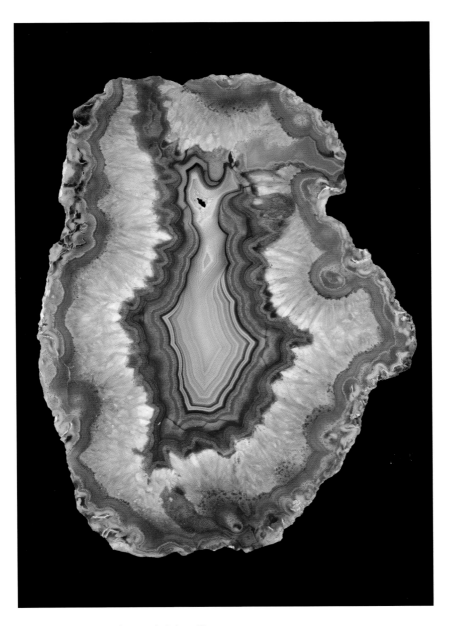

Above Laguna agates can become fairly large. This
example has a floating centre, a pod of banded agate
that is suspended in a matrix of euhedral quartz. Length
of specimen approximately 15 cm (6 in).

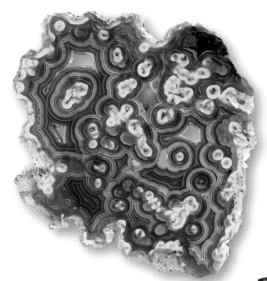

Left The many tubes in this Laguna agate formed around needles of goethite, a hydrated oxide of iron. Diameter approximately 10 cm (4 in).

Below The most interesting patterns of the agate are often near the periphery of the nodule. The eye in the lower right is caused by a pit on the surface of the nodule. Diameter approximately 12.5 cm (5 in).

Ojo Laguna is a railroad siding that serves several large cattle ranches and is situated about 120 km (75 miles) south of Ciudad Juarez along the Rio Grande River on the USA-Mexico border. The Laguna agates from here are probably the favourite of agate collectors throughout the world, and many fine museum and private collections in Britain, France, Germany, the USA and Japan boast specimens from this area. The agates from here have been mined from a geological unit called Ranch el Agate Andesite. Because these agates are geologically young, they contain many kinds of inclusions that have not been destroyed by the effects of age, weathering and transportation. Mining and collecting activity was curtailed during the Second World War, and they received no attention until the late 1940s when

Above Tiny but highly reflective crystals of drusy (minute crystals) quartz provide a great deal of eye appeal to many Moctezuma agates. Diameter of agate approximately 3.7 cm (1¹/₂ in).

they again appeared on the markets in the USA. Many Laguna agates demonstrate a phenomenon called 'shadow' or 'parallax'. This is caused by alternating, fine, highly transparent and opaque bands that create a shadow on the opaque surfaces that are adjacent to the transparent bands. This phenomenon can be duplicated by stacking alternating sheets of clear, colourless acetate, and white or pale coloured paper.

Estación Moctezuma is another railroad siding about 120 km (75 miles) south of Ciudad Juarez, in Chihuahua, Mexico, that has lent its name to a very colourful agate – Moctezuma agates. They are commonly mislabelled as 'Montezuma' agates. Similar agates from the same region in Chihuahua, Mexico, have been called Barney Ranch agates or Varney Ranch Agates, but these names are rarely used anymore. Marine sedimentary rocks of Cretaceous age (65-142 million years old) and volcanic rocks of Tertiary age (1.8-65 million years old) make up the landscape. Moctezuma agates are characteristically small compared to those agates from Ojo Laguna or Rancho Coyamito, most specimens ranging in diameter from about 25 to 50 mm (1 to 2 in) making them popular for jewellery making. Most Moctezuma agates have been weathered out of their source rock, but their exteriors show no signs of having been worn or abraded. They appear to form a lag of resistant fragments that are too large to be blown away by wind or washed away by the

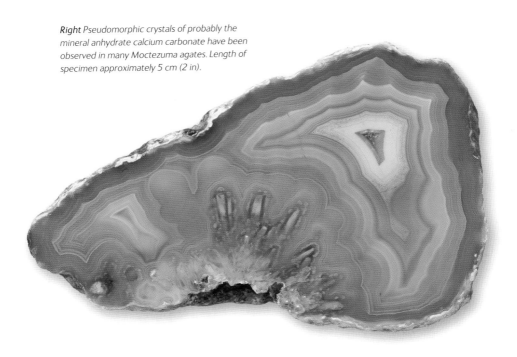

Right Pseudomorphic crystals of probably the mineral anhydrate calcium carbonate have been observed in many Moctezuma agates. Length of specimen approximately 5 cm (2 in).

meagre rainfall in the area. Because Moctezuma agates are from a relatively young geological setting, many still contain adhering pieces of host rock as well as many different kinds of structures and inclusions. Most of the Moctezuma agates are weathered on the outside to white, clay-sized particles, but they show no evidence of having been stream abraded. They are commonly opaque, but the colours may range from dull pastel shades to brilliant hues. Almost every colour but blue or green have been observed in these agates. In addition to their bright colours, Moctezuma agates contain a large variety of inclusions such as crystal pseudomorphs, stalk aggregates, plumes, etc.

Gallego agates derive their names from the Sierra del Gallego Mountains, a range that has produced many varieties of agate, including most of the above varieties. These are generally a light purple or lavender colour, and many of the nodules are hollow and have attractive crystal-lined centres. Many of the Gallego agate nodules show identical sequences of banding. Gallego agates generally have few, if any, inclusions. Most of these agates are relatively small, less than 5 cm (2 in) in diameter. Because most of them have hollow, crystal-lined interiors, they are not commonly used as jewellery, but many are very attractive cabinet specimens.

Agua Nueva agates derive their name from Rancho Agua Nueva, and these include both vein and nodular agates. Some of these agates become quite large, up to 30 cm (1 ft) in diameter. Agua Nueva agates rarely contain inclusions other than decomposed matrix rock and membranous cristobalite or chlorite, but one specimen shows very rare disc-like inclusions that were first recorded on a Scottish agate by Heddle (1901), see p. 26.

Above *Agua Nueva agates are named for Agua Nueva Ranch, in northern Chihuahua, Mexico. They are usually mined from relatively unweathered rock and the surfaces of the nodules show many features that are lost on more weathered agates. Their bands are generally sharp and brilliantly coloured and many contain mossy inclusions. Height of specimen approximately 15 cm (6 in).*

Left *Pink, violet, and yellow shades are common in Agua Nueva agates. Length of specimen approximately 13.7 cm (5¹/₂ in).*

Above Coyamito agates often come in brilliant shades of red and yellow. Specimen approximately 7.6 cm (3 in) long.

Coyamito agates derive their name from Rancho Coyamito, a large cattle operation that is quite close to Ojo Laguna. These agates are amygdaloidal, and although they are called Coyamito agates, they commonly come from commercial shipments of agates from various sites that have been mixed together. These agates are very colourful and many contain various inclusions such as crystal pseudomorphs, plumes and zeolites. Some of these agates provide outstanding examples of iris, a phenomenon that is caused by the bands working as a diffraction grating, breaking white light down into its component colours, red, yellow and blue.

Below *Violet and yellow Coyamito agate with large dilation at left of picture. Length of specimen approximately 6.3 cm (2¹/₂ in).*

Above Bubble lace agate is so called because it has a widely banded texture that resembles bubbles forming on the bottom of a boiling pot. Length of specimen approximately 21 cm (8 in).

Below This piece of lace agate has the typical wild, lacy pattern on the left side and a finely banded pattern on the right side. Length of specimen approximately 25.4 cm (10 in).

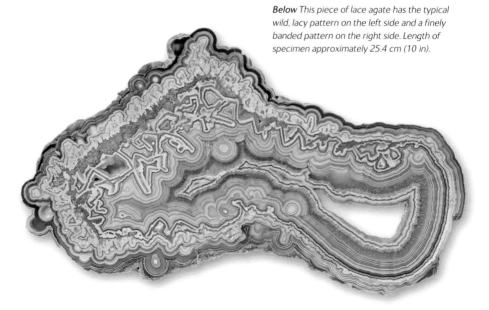

Most iris agates are thought of as being colourless, but many brightly coloured, translucent Coyamito agates produce a masked iris in which, for example, a red agate will absorb the red of the spectrum while the blue and yellow components are visible. Iris is commonly observed in very thin slices of polished agate, 1.5 mm ($^1/_{16}$ in), but some examples as thick as 6 mm ($^1/_4$ in) may show outstanding iris. Iris is easily observed in transmitted light, with light passing through the back of the specimen. As the viewer raises and lowers the agate at about arms' length the spectral colours will change. Incandescent light or sunlight are the best lights for viewing iris. You should exercise extreme caution when viewing iris agates by sunlight, however, as eye damage could result. Many newer incandescent lamps have eliminated much of the red and yellow spectral colours from their light, and iris agates frequently appear dull in these lights, just as they do with fluorescent lights.

In the state of Chihuahua, crazy lace or lace agates are mostly vein agates found in marine sedimentary rocks that have been mapped as late Cretaceous age (65-142 million years old) and which formed along fault or joint planes in the dark grey limestone. (Faults are fractures in rock where there has been movement along adjacent blocks, whereas joints are fractures in rocks where there has been no movement along adjacent blocks.) There are many variations in the material that is called lace agate. These fracture planes provide excellent conduits for mineral-rich solutions to deposit silica that forms the agate. Red and white lace agate is sometimes offered as Quetzalcoatl agate, named after the Aztec deity. Dogtooth lace agate is characterised by containing agate pseudomorphs of calcite crystals. The crystal pseudomorphs are relatively large, and most of these agates find their niche as cabinet specimens. Cactus lace agate has many fine zeolite crystal inclusions that resemble cactus needles, and bubble lace agate has a bubbly appearance.

Mexican flame agate contains red, yellow and orange-red parallel oriented streaks, perhaps of iron oxides, that resemble forest, prairie or grass fires. The finest flame agates have a very clear, translucent matrix that gives some depth to the specimen. Bird of Paradise agate is very similar to Mexican flame agate, except the streaks of colour are randomly oriented, and Bird of Paradise agate commonly contains additional pastel shades such as pink or peach-coloured streaks.

Above *Agate with fish pattern from Uruguay.*

South America

Uruguay

The first major source of agates from South America was from Uruguay. These were discovered by German immigrants in around 1830 and quickly became a source for the agate cutting industry in Idar-Oberstein, Germany. Some agates are still mined in Uruguay, but sources in Brazil proved to be of greater importance in the late 19th and 20th centuries.

Brazil

There are at least 15 major agate-producing areas in Brazil. Twelve of these areas are situated in the five southeastern most states of the country, Rio Grande do Sul immediately north of Uruguay being the best known. Additional important agate producing areas are in the states of Santa Catarina, Paraná, São Paolo, and Minas Gerais. Two agate producing areas are in the state of Bahia in east central Brazil, and one other agate producing area is in the state of Roraima in Brazil's extreme northwest. These agates were discovered in Brazil shortly after the finds in Uruguay. In the late 1920s, a geologist named B.V. Freyberg, under the employ of the Geological Survey of Argentina, produced a detailed study that began with the

Above *Agate-producing areas in Brazil.*

Uruguay agate deposits, but covered mostly the Brazilian agate deposits. In 1974 a geologist named L.E. de Mattos produced an economic geology study on the agate-mining industry in Brazil, and he gave locality data for some of the named agate varieties. Some of these variety names have been altered by dealers or collectors in attempts to enhance the saleability or value of some of these gems. For example, Parana agates have been called Piranha agates after the voracious cannibal fish. Each area produced a distinctive kind of agate. The area around Espumoso is known for outstanding landscape agates. Carazinho agates are commonly finely banded and have colour patterns arranged perpendicular to the bands. Parana agates commonly have very bright colours that are in large splotches and produce some outstanding large cabochon gems.

Left Agate from Rio Grande do Sul, Brazil, with large eye-like bands and dilation in escape tube on left side. Diameter approximately 12.5 cm (5 in).

Current laws in Brazil prohibit the exportation of unfinished gemstones, including agate. Before these laws went into effect, agate cutters and fanciers outside of Brazil could purchase government sealed kilo boxes that contained 100 kg of rough agate. The agate was packed into the boxes by the distributors and stones from various areas were commonly mixed together to fill up a box.

The chief products from Brazilian agates are ornaments and jewellery, but some is still used for scientific purposes such as bearings, and mortar and pestle sets. No current production figures for Brazilian agates are available, but in the early 1970s the annual output of agate increased in tonnage each year, but its value made a declining percentage of Brazil's gem exports, the value of the agate being overshadowed by more precious stones. During the early 1970s the export value of manufactured items of Brazilian agate was far less than the value of the rough stones. This is because most cutters prefer to work with the untouched rough. In the late 1980s and early 1990s, the value of rough Brazilian agate fell to very low levels, and buyers could obtain choice material for as little as $0.15 US per pound, and large, polished specimens for $10.00 US or less. Since the exportation laws have been in effect, the prices of Brazilian agate have recovered somewhat.

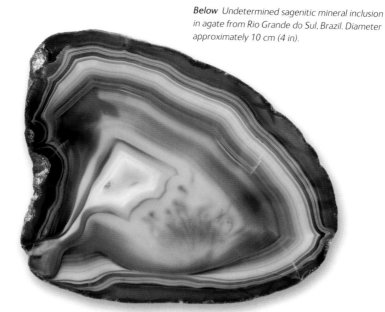

Below *Undetermined sagenitic mineral inclusion in agate from Rio Grande do Sul, Brazil. Diameter approximately 10 cm (4 in).*

The agates from Brazil have been mined mostly from decomposed volcanic ash and basalt of Late Permian age (248-275 million years old). Agates are abundant in many areas of Brazil, but especially in Rio Grande do Sul where bedrock maps show how extensive the deposits of volcanic rocks are. Many of the mines amount to little more than ploughed fields from which the loose agate nodules are collected.

Many Brazilian agates are very colourful stones with some of the most unusual structures and inclusions that an agate student or collector could imagine including tubes, eyes, sagenite, plumes, crystal pseudomorphs, and stalk agates. Strangely, the agate miners do not especially care for these stones. The miners prefer what have been called industrial grade agates, shades of pale yellow or grey, or colourless agates. These stones are readily porous and permeable, and take artificial dyes very readily, so most of the market potential for Brazilian agate is in the large, dyed objects such as bookends, paper weights, pen bases and spheres (see p. 157).

Nodules of Brazilian agates can be quite large. Some specimens may reach 0.9 m (3 ft) in diameter, and weigh over 120 kg (300 pounds), and specimens 0.3 m (1 ft) in diameter are not uncommon. They are probably, on average, the largest agates known.

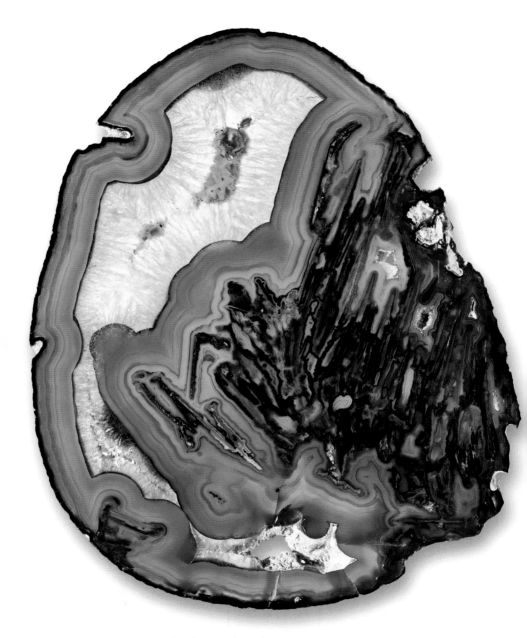

Above Brazilian agate with undetermined crystal
pseudomorph inclusions. Height of agate
approximately 15 cm (6 in).

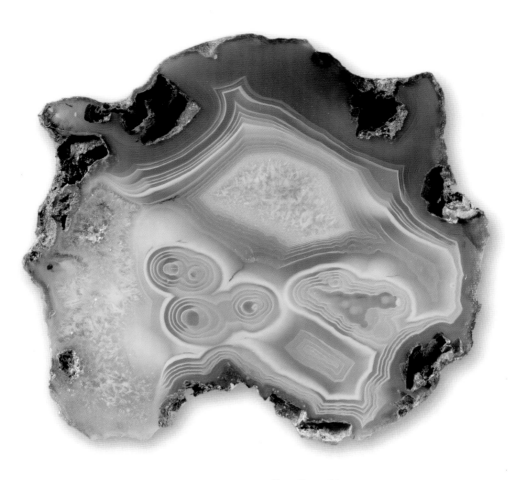

Above *Some of the newly found agates from Chile retain a fine blue colour even when cut into thin sections. Specimen approximately 5 x 5 cm (2 x 2 in) in diameter.*

Chile

Plume and banded agates have recently been found in volcanic agates in Chile. The ages of the rocks in which these agate have been found have not yet been determined by geologists. The Chilean agates are characterised by being extremely tough. The original miners report suggested that they were of a hardness of 9 on the Mohs scale. The toughness, caused by the twisting together of the microcrystals that make up the agate, in this case gave the illusion of hardness to the cutters.

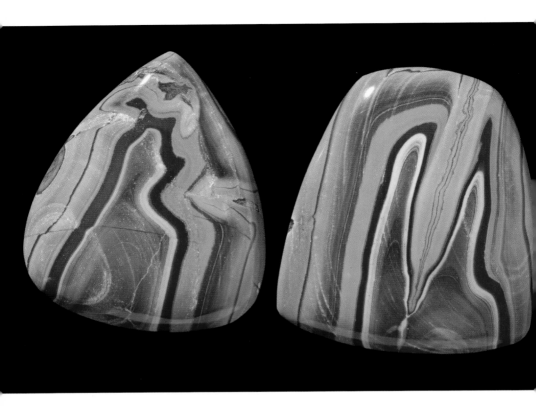

Above Rhyolite from Peru is commonly sold as agate, and its banded structure suggests to many novices that it is really agate. Largest cabochon approximately 5 cm (2 in) high

Peru

Peru is best known for rich deposits of copper minerals and guano, an important phosphate based fertilizer, but not well known for agate deposits. However, in the early 2000s, an non-banded, light to medium green stained chalcedony arrived in the gem trade. The green stains possibly result from impurities of copper minerals such as malachite or atacamite. These stones may be from the Cerro de Pasco area, a rich copper producing district but no records are available at this time. There is a variety of finely banded rhyolite from Peru that has been marketed as agate. It is an attractive stone, but it is not agate in any sense of the word.

Argentina

Argentina has become an important source of agates since about the mid 1980s. All of the Argentine sources are in remote areas and the cost of recovering the agates is very high, but the outstanding colours and patterns make these stones highly desirable and they are popular with many collectors. In addition to having many outstanding agates, Argentina is also the source of some excellent examples of agatised wood.

Condor agates were the first of the new generation of Argentine agates to reach the world markets. These are commonly amygdaloidal banded or fortification agates with especially bright red and yellow colours that have been found in the

Above Many Condor agates have bright red and yellow, highly contrasted bands, and many are relatively free of flaws and fractures. Height approximately 7.5 cm (3 in).

Above Typical example of crater agate from Argentina. Length approximately 12.5 cm (5 in).

Below Stream or wave tumbled agates from Tierra del Fuego, Argentina, are of currently unknown origins. Largest stone approximately 6 x 10 cm (1$\frac{1}{2}$ x 2 in).

Mendoza province of Argentina. Some contain mossy and sagenitic inclusions.

Another popular Argentine agate is called crater agate. Information provided by a miner of gems from this area suggests that they formed in rhyolitic rocks of Jurassic age (142-206 million years old) in the Patagonia area of Argentina. Many crater agates are black with prominent red bands near the centre of the commonly hollow nodules. Puma agates, according to information supplied by the miner, are of marine sedimentary origin, and appear to be pseudomorphic. They have been found in the Andes in Patagonia. Microscopic examination shows what appear to be septa and dissepiments in corals, but recrystallisation of the rock is so complete that it is difficult, if not impossible, to determine the species of coral that became agatised here.

Recent stream and beach deposits in Patagonia and Tierra del Fuego in southern Argentina have yielded examples of small, stream and wave tumbled agates that are from currently unknown sources. These agates are usually small, 2.5-5 cm (1-2 in) in diameter, and many are complete nodules. Some have been observed for sale in shows or catalogues, and examples have been collected by graduate students and travellers to these areas. Some of these agates have recently been offered by show dealers as Black River agates.

Africa

Botswana

Botswana agates are amygdaloidal agates that have been found in great profusion in basaltic rocks of Permian age (248-290 million years old) in the Karoo Series of South Africa. Except for colour, many of these agates resemble the Lake Superior agates of the USA mid-continent, and they have sometimes been referred to as African Lakers. The similarity of these diverse agates may reflect similar geological conditions at two different times and in two distant locations. In truth, Botswana agates are more closely related to the Queensland agates of northeastern Australia.

Botswana agates commonly come in shades of deep purple, medium to dark-greys, to black with sharply contrasting white bands. Deeply weathered specimens tend to have faded to shades of pink and light grey. Several kinds of structures and

Right This fine cabinet specimen of Botswana agate shows sharply contrasting bands as well as inclusions of green magnesian chamosite. Specimen approximately 10 cm (4 in) long.

inclusions have been observed in Botswana agates, and these include sagenitic inclusions, rare plumes and stalk aggregates. Uruguay banding is very rare in these agates. They are commonly small, being 2.5-5 cm (1-2 in) in diameter, but some examples up to 15 cm (6 in) have been observed. The small agates have become very popular especially with amateur gem cutters as they show very exquisite and delicate patterns with sharply contrasting bands.

South Africa

South Africa has produced a very interesting kind of agate that is a by-product of alluvial diamond mining operations. These agates are referred to as 'diamond diggings' agates' as they have been collected from depleted alluvial diamond

Left The black stripes are cryptocrystalline agate and the white are quartz. Size of slice approximately 10 x 10 cm (4 x 4 in).

deposits in the northwestern part of South Africa. These agates were known to geologists and gem cutters as early as 1861, and the term 'Bantom' was used to describe these stones in 1965, referring to the bands around the stones. They are normally fortification agates with translucent grey and opaque black bands on cut surfaces, but yellow and dark grey on weathered surfaces.

Transvaal in the northeastern part of South Africa is home to some very brightly coloured, small amygdaloidal agates. Their source has not yet been determined, but they may have formed in volcanic and basaltic rocks of Permian age (248-290 million years old).

Zebra lace agate derives its name for the alternating black and white strips. This agate is unusual inasmuch as the black stripes are cryptocrystalline agate, whereas the white, stripes are quartz crystals. Similar agates have also been found with dark green and white and red and white stripes, in India as well as South Africa.

Malawi

Malawi has produced some of the most beautiful agates to have been found on the African continent. Most of these stones are brilliant shades of red or orange with brightly contrasting pure white bands. Some have pink bands in a pale blue groundmass, see p. 29. Most of these agates probably formed in volcanic or basaltic rocks of Permian age (248-290 million years old). Most examples appear to be amygdaloidal rather than thunder eggs. These agates arrived in the trade in the late 1940s and early 1950s.

Namibia

Namibia is known for blue lace agate, a delicately coloured blue gem that retains its colour in very thin slices. This gem first appeared in the middle 1970s, with the source being listed as South Africa, but advertisements appearing in the late 1980s stated Southwest Africa as the source. These stones were likely first reported as blue lace chalcedony, in which case this name should have priority. Frazier and Frazier (1988) called these stones African blue lace agate, and listed Namibia as the

Above *Attractive blue lace agates still retain their fine light blue colour in very thin slices.*

source, and indicated that South Africa, Southwest Africa and the Kalahari Desert were also sources. These agates formed in dolomite associated with an igneous dolerite according to geologists at the University of Stellenbosch in South Africa.

Egypt

Egyptian jasper is a term used to describe jasper that has coloured bands which do not form the same way as they do in agates. The bands are irregular, vary greatly in thickness, may be discontinuous, and are frequently truncated by another set of bands that are oriented in another direction. Such bands are usually light and dark shades of the same colours, for example, light tan and dark brown bands or light grey and dark grey bands. This kind of structure is common to many picture jaspers throughout the world.

Zimbabwe

Zimbabwe has produced agates from several areas including Bembezwane, Glenaroch Road and Lake Cariba. Few, if any, of these agates have reached collectors or museums in other countries. Most of the agates described are banded agates. Chrome chalcedony is one form of agate that has come out of Zimbabwe; it is a dark green material that may be an agatised serpentine.

Opposite and this page Ocean Jasper®
from Madagascar is a relatively new
material to appear on the market.
Although it has been known for many
years, the sources were not discovered
until the late 1990s. Largest cabochon
(opposite) approximately 7.6 cm (3 in)
long.

Madagascar

Madagascar has not generally been known for agates until recent years. In the late
1990s a very colourful material called Ocean Jasper® appeared on the world gem
market. According to the materials provided by the name owners, this gem had been
known from loose pieces seen in bazaars and shops in Madagascar, but the source
was unknown. The name Ocean Jasper® was derived from the proximity of the
deposit to the Indian Ocean. The stone is characterised by having colourful and
contrasting spherulites and pods of agate distributed in a granular matrix. This stone
has been a very popular material with jewellery makers and sphere makers. Some
fine examples of agatised wood and banded agates have also been found in
Madagascar, but little is currently known about their sources.

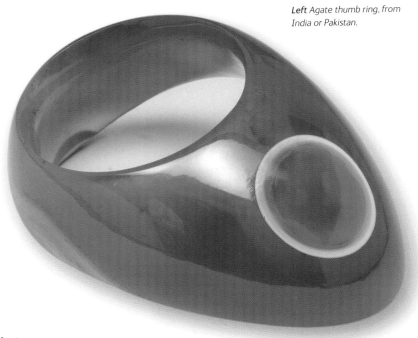

Asia

India

India has produced a large variety of agates, and an active agate cutting industry existed in the state of Gujarat in northwestern India near the Narbada River which flows into the bay of Cambay. The agate industry there probably pre-dates the Idar-Oberstein agate industry by several hundred years as research shows it was functioning as early as AD1000. The agates from this area have been mined from stream deposits of Pliocene age (1.8-5 million years old) that have been derived from erosion of much older rocks called the Deccan Traps of Late Cretaceous age (65-100 million years old). The Deccan Traps include extensive flood basalts that cover over 500,000 km² (200,000 miles²), produced by enormous lava flows. Some scientists think that the gases released during these eruptions may have altered the atmosphere and led to the demise of the dinosaurs, and some scientists think that the asteroid collision at the end of the Cretaceous may have triggered these massive eruptions. Many of the kinds of agates found in this area have been observed in place in the Deccan flood basalts of very Late Cretaceous age.

Right Fine example of agate with mossy inclusions in stalk aggregates from northwestern India. Diameter approximately 5 cm (2 in).

Written descriptions of the Indian agate mining and lapidary industry are among some of the oldest in the known literature. Varthema published accounts of Indian agates as early as AD1510, and Tavernier in AD1676. In 1776, Collini published an account of these agates being found in basalt. These early publications were in Latin, French and German respectively, and the first English language paper dealing with the Indian agates was probably written by Dr Copland of the Bombay Medical Service in 1819. He recorded that the mining procedures were very wasteful. Pits about 1.3 x 1.3 m (4 x 4 ft) were dug to about 9 m (30 ft) deep. Horizontal shafts were extended from these. The mines were not worked during the monsoon season, and previously dug pits were allowed to cave in, requiring a new pit be opened each year. He suggested that much material remained un-mined. Stones may have been as large as 10 cm (4 in) in diameter. Many stones were colourless, possibly due to the low oxidation states of iron compounds, when they reached the surface. The newly mined stones were carried from the mines to the city of Limodra for safekeeping after each day's work. Newly mined stones were exposed

Opposite India dendritic agates. These stones
have also commonly been known as mocha
stones in allusion to similar stones that have been
historically produced from Yemen. Largest stone
approximately 6.7 cm (2¹/₂ in) high.

to sunlight for several months, and were turned over every 15 days. About a month before the monsoon season came, the stones were fired – heat-treated to produce the desired carnelian colour, the colour in greatest demand. The stones were placed in clay pots over a fire of goat dung fuel that was set alight in the evening and allowed to burn through the night. Most of the fuel was consumed by morning and the stones were allowed several hours to cool. The edges of the stones were chipped and those that did not change colour to carnelian were re-fired. Those stones that reached the desired colours were sent to the lapidary shops to be turned into gemstones and ornamental objects such as bowls and knife handles.

In addition to the carnelian agate mentioned above, India has been the home of some very fine moss agates. One kind of moss agate consists of a green chlorite mineral that is enclosed in a nearly transparent, colourless, to opaque, white groundmass. Both varieties have been used for jewellery, and sufficiently large pieces have been recovered that allow ornamental objects such as bowls to be crafted from them.

A second kind of moss agate from India consists of dendrites of various oxides of manganese and iron in a nearly white to slightly yellowish groundmass. This agate very closely resembles the fine dendritic agates from Yemen and has been sold as mocha stone, the name given to the Yemen material. This material has been recorded as being from the state of Rajpipla, but neither Dr Copland nor later investigators recorded seeing it in their writings. In addition to carnelian and moss agates, India has produced some outstanding examples of jaspers and bloodstones.

Yemen

Yemen is of both great historic and geological importance as a producer of agate, and agate beads dating back as far as 4000BC have been found in Yemen. The agates probably formed in volcanic-ash tuff deposits of Late Oligocene and Early Miocene age (10-40 million years old). These agates have been called mocha stones, after the city of Al Mukha (Mocha) on the Red Sea.

Interestingly, agate from Yemen may be instrumental in preserving existing populations of the endangered African black rhinoceros, *Dicerocornis bicornis*. The

World Wildlife Fund has suggested that the horns of these animals were very popular in Yemen and other neighbouring countries for dagger handles in the 1970s, so endangering the black rhinoceros populations Locally mined agate from Yemen became an acceptable and desirable substitute for rhinoceros horn in the mid-1990s, thus easing the demand for these horns.

Iran

The Howe-z-Mirza Mountain in the Khur area of Iran is an important agate-producing area. This area has produced some very fine examples of pale to deep blue agate from volcanic rocks of Early Eocene age (40-58 million years old). Plume, moss, sagenitic, flame and pompon (a kind of sagenitic agate resembling the flower, a type of chrysanthemum) agates have all been found, as well as mineral inclusions such as calcite, aragonite, dolomite and siderite to mention but a few.

Below The Khur area of Iran is a very important source of agates with many different kinds of structures and inclusions. This light blue agate is approximately 5.1 cm (2 in) in diameter.

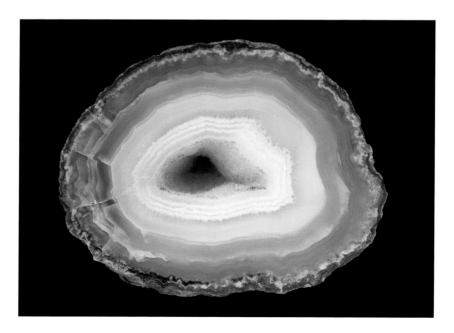

Mongolia

Mongolia is a rich source of agates but has not received much attention because of its geographical isolation. Some very beautiful thunder egg and amygdaloidal agates have been found there.

China

China has become an important source for agate jewellery, ornaments and utensils, but little is known about their place of origin. Much of this material has been seen in gem and mineral shops and shows, and some dealers have stated that the material has come from the Changjiang River near Shanghai. Some pieces appear to have been artificially coloured.

Above Snuff bottles, 1780–1909, China.

Below An agate from Agate Creek. This volcanic region in northern Queensland, Australia, is a fine source for agate-hunters.

Australia

Queensland agates were first recorded in a 1900 report that dealt with gold resources in this remote region of northeastern Australia. These agates formed in basaltic lava flows of Late Permian age (248-275 million years ago). Queensland agates were generally not known outside of Australia until after the Second World War. During the war, some 400 tonnes of these agates were used for military purposes, many for turn and bank indicators in aircraft or bearings for precision instruments. Most of the Queensland agates that are found in rock shops are small, being 5 cm (2 in) or less in diameter, but some large nodules of up to 25 cm

(10 in) have been observed. Queensland agates are often characterised by their rich colouring, and by several hues and shades that are usually not observed in agates from other parts of the world. Inclusions of various kinds are not common in Queensland agates, although a few specimens with crystal pseudomorph inclusions have been observed. Uruguay or level banding is very common in these stones.

There are many other beautiful agates that have been found in Australia. The geographic remoteness and cost of transporting these stones to other parts of the world is prohibitive to most dealers or collectors outside of the continent. As a consequence, most of the Australian agates seen in other parts of the world are usually small.

Several kinds of thunder egg agates have been found in Queensland from the Cedar Creek deposits at Tamborine Mountain, Thunderbird Park, a popular tourist

Below *Agate-producing areas in Australia and Tasmania.*

Opposite Uruguay banding is common in Queensland agates from Australia, and it usually contrasts sharply with the wall's lining bands. Diameter approximately 3.7 cm (1¹/₂ in).

Below This shade of light greenish yellow is almost never seen in any kind of agate but the Queensland agate. Diameter approximately 5 cm (2 in).

resort and collecting area. These gems have been called Mount Tamborine thunder eggs, Cedar Creek thunder eggs, and Doon-Doon thunder eggs, and they are a favourite stone among many Australian agate collectors. It has been suggested that the thunder eggs were of Early Cretaceous age (100-142 million years old), and a model was developed by Kay (1960) to show their formation where colloidal silica accumulated about a suitable nucleus such as a crystal. The silica eventually crystallised in a spherical mass and went into a distended phase where volatile compounds caused expansion, leaving a hollow that eventually filled with agate.

Rock collecting is commonly referred to as 'fossicking' in both Tasmania and Australia, and Mineral Resources Tasmania has prepared several short brochures that cover some of the agate collecting areas there. The Lune River in Tasmania is the source of some agates that have commonly been called Lune cone onyxes. Most of the specimens from there have the Uruguay or flat, layered bands. Many of these tend to be sub-triangular in cross section, and many contain interesting structures such as stalk aggregates and mineral inclusions. Other agate producing areas in Tasmania are at Coal Hill near Hastings, Tunnel Marsh near Tarraleah, and Weymouth.

Opposite Agate from New South Wales, Australia.

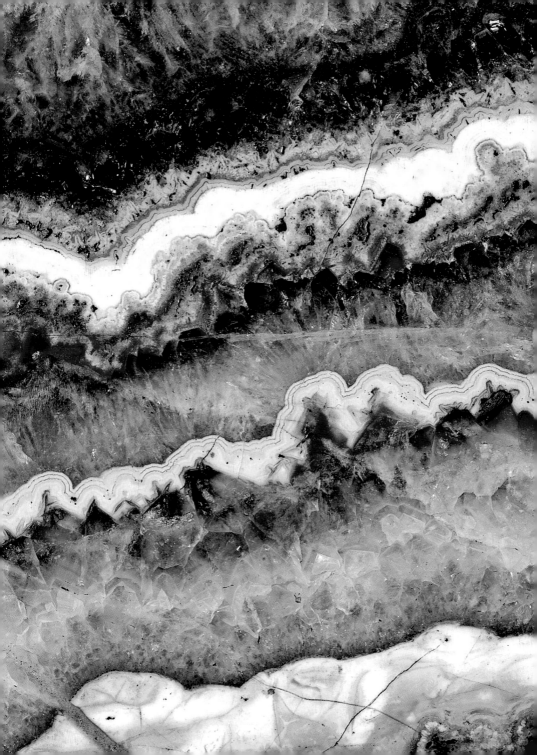

Lapidary

Techniques used by the lapidary

Sawing

Saws are the first in a series of tools the lapidary or cutter uses in order to shape a piece of agate into the desired object. The most primitive saws were no more than a piece of wire strung to a bow. The bow was moved back and forth, and quartz sand and a small amount of water to create a slurry were poured onto the wire. Eventually, the hard abrasive wore through the stone and the agate was cut into smaller pieces. Later, the bow and wire saw was replaced with the rotary (or rotating wheel) saw, consisting of a rotating blade that passes through a slurry of sand and water which acts as an abrasive to cut through the agate. Small saws were initially powered by a foot treadle, while larger saws were commonly powered by water, at least until the development of the electric motor in the late 1800s.

Opposite Agate, unknown locality.

Right Trimming a slice of agate with an oil-cooled diamond saw.

Another advancement in lapidary technology came with the development of Carborundum™ (silicon carbide) and aluminium oxide as abrasives. The much harder and crystalline silicon carbide cuts through the hard agate faster than the quartz sand.

Diamond saws were developed around 1930. The earliest diamond blades consisted of finely crushed diamond powder being inserted in a cut along the edge of a copper or brass disc. The diamond powder was inserted into the disc and the cut was hammered shut. These processes were eventually automated and soft steel became the selected metal for diamond blades, which became relatively inexpensive and the choice of most lapidaries. The earlier diamond blades required a mineral oil coolant that was expensive, odoriferous and toxic to some individuals. The late 1900s saw rapid advances in diamond technology and the old oil-cooled blades quickly gave way to blades that function even better, and use only a water coolant.

Grinding

The lapidary removes as much of the excess matrix material from the stone as possible with the diamond saw. Sawing not only speeds up the operation but is also more economical than using a grinder to remove all the excess material.

The grit size of a grinder will tell you how fast the grinder will abrade the subject stone. Grit size indicates the relative coarseness of the grit. On most grinders this is determined by the number of particles of abrasive per inch if the particles were aligned side by side. The lower the number the coarser the grit and more abrasive it is. The higher the number, the finer the grit on the grinding wheel. More recently, most diamond grit compounds are measured in microns (1 micron = 1/1000 millimetres); thus a 5 micron grit would be much finer than a 500 micron grit.

The earliest lapidaries probably used only loose sand and hardwood sticks to do much of the shaping of their stones. Pieces of quartzite, a metamorphosed sandstone, were also likely tools for the primitive lapidary shop. It has not been established when the first rotating grinding wheels came into use, but they were a fixture in many lapidary shops by the late 1300s. The earliest grinding wheels were large, up to 3 m (10 ft) in diameter, made of very hard sandstone or quartzite, and were powered by flowing water.

By the late 1700s or early 1800s lapidary hobby units came into production. These small units consisted of a flat, cast iron lap that was turned by a hand crank. The lapidary attached the stone to a dop stick and held that in one hand and turned the crank with the other. A slurry of sand and water acted as a grinding agent.

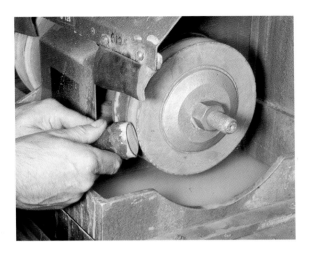

Above Shaping a cabochon on a coarse, water-cooled grinder.

With the development of silicon carbide and synthetic aluminium oxide, grinding wheels were made from these compounds. Silicon carbide became the choice for lapidary grinding wheels, whereas aluminium oxide proved to be much better for work with metal. The silicon carbide grinding wheels were a great improvement over all previous tools, but they were expensive, messy and needed to be constantly refaced to prevent rough grinding.

Diamond grinding wheels for lapidary use first appeared in the early 1960s, and they did not receive a favourable review. They were expensive and did not wear well; a grinder could be denuded of diamond in a fairly short time. The manufacturers did not give up on their product and the technology improved quickly. By the middle of the 1970s many lapidaries had switched from silicon carbide to diamond grinders. Now, most lapidaries are committed to diamond tools. In addition to the vertically oriented diamond grinders, horizontally oriented diamond units have become available at lower cost. The development of synthetic diamond proved to be beneficial for the lapidary tool industry, as the synthetics were equally hard and much less expensive than the natural industrial diamonds.

In spite of the fact that diamond grinders have increased in popularity, they will never completely replace the silicon carbide grinder, and the best lapidary shops still have silicon carbide units available. The creative lapidary will have hundreds of different tools at his or her disposal.

Sanding

Sanding the shaped stone is the next step the lapidary follows on the way to a smoothly contoured stone. After the stone is ground, its cross-section will be roughly circular or parabolic, but the contour will be interrupted by small ridges and flat facets that must be removed to give the stone an attractive surface. Here the lapidary uses soft and pliable belts or disks that are coated with either silicon carbide, aluminium oxide or diamond-coated sandpaper. The coarseness of silicon carbide coated sandpaper is graded in exactly the same way as the coarseness of grinding wheels. The pliable backing of the sandpaper during this operation gives the lapidary what is needed to remove the facets, ridges and flat spots on the ground stone. The lapidary uses several, continually finer grades of sandpaper or diamond compound to achieve a semi-polish on the stone. On flat laps of around 25 cm (10 in) diameters, 600 grit wet and dry paper gives a very fine finish, provided that loose 600 grit mixed with detergent is applied sparingly at the same time. This should leave a scratch free finish, often with a dull polish.

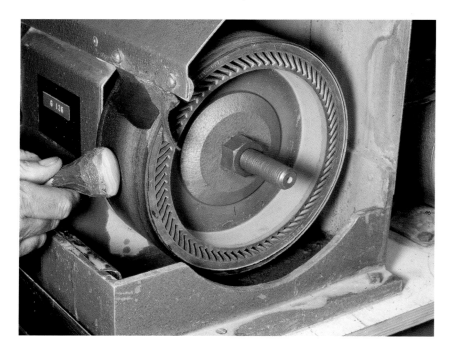

Above Sanding a cabochon on a resinous belt
that is charged with pulverized diamond.

Above Polishing a cabochon on a soft leather pad
using chromium oxide as a polishing compound.

Polishing

A final polish can be achieved by using extremely fine sandpaper (14,000 or 50,000 grit) or by using a polishing compound such as cerium oxide, titanium oxide or tin (stannic) oxide. Some of these polishing compounds are sold under various trade names, but they are usually one of the above generic compounds.

The polishing compound is applied with water to a wheel or disc of leather, felt, hardwood or some newer synthetic material. If the stone is properly sanded, the polish will be achieved in a few seconds. A superb polish can be obtained by using a flat lap with a 25 cm (10 in) hard felt wheel, run at 1500 rpm, using cerium oxide powder and sufficient water that allows the stone to become hot and promote 'beale flow' (see below). Considerable finger and hand strength is necessary but with practice will always give exemplary results.

These polishing compounds are all much softer than the agate or most other stones they will polish. There are several theories as to why a softer compound will polish a much harder stone. One such theory is that the compound causes the water to superheat at the interface with the stone and the disc or wheel, thus, melting a very thin layer at the surface of the stone (beale flow). Another theory holds that the physical structure of the polishing compound is altered at the interface between the stone and disc, causing the compound to become very hard.

Imitations and forgeries of agates

During the Victorian era, mourning jewellery was commonly worn in memory of the deceased. Most of this jewellery was made of black jet from near Whitby in Yorkshire, England. Some unusual pieces included assembled doublets made from two layers of a transparent stone, probably quartz, with short pieces of the hair of the deceased cemented between the two quartz layers, resembling moss agate. This kind of stone appeared as early as post-revolutionary times in France, but was not necessarily worn as mourning jewellery then.

Several kinds of obvious forgeries of agates emerged in the late 1900s and early 2000s. Large specimens of Lake Superior agates from the North American mid-continent have commanded high premiums for many years. This fact led unscrupulous dealers to heat-treat very common, large, whole, colourless agate nodules from Brazil or Uruguay to produce a reddish to reddish-brown surface on the nodule, resembling the Lake Superior agates. Some of these forgeries were sufficiently good enough to fool many dealers and even knowledgable collectors. Such forgeries can usually be detected by magnification of the outer surface. The Brazilian and Uruguayan agates are stream deposited and have little 'half-moon' fracture marks all over the outer surface, whereas the glacially deposited Lake Superior agates show few, if any, such marks. Although large Lake Superior agates are not commonly cut open, those forgeries that were sliced open showed that the colour was only 'skin deep' and penetrated into only about 2 mm (1/8 in) beneath the surface.

Polished half nodules of high-quality agates, especially those from northern Mexico, command high premiums, and a new kind of forgery turned up in 2002. This was a very thin slice of banded agate that had been cemented to a block of rhyolite. This gave the effect of a rather large half nodule. The forgery was made even more believable because the surface of the nodule had been polished with green chrome oxide. The chrome oxide residue left the impression of green celadonite on the surface of the faked nodule. Had not the agate slice fallen off this forgery, it might still remain undetected. The other striking feature of the fake half nodule was that it was not of a particularly high quality stone. One needs to exercise extreme caution even when buying polished half nodules.

Plume and dendritic agate cabochons with particularly attractive designs also command high premiums. A kind of forgery that came about in the late 1980s and early 1990s probably owed its emergence to advances in mathematics and computer technology. Textbooks on higher mathematics such as fractal geometry provided many images of plume and dendritic patterns that had been computer generated. Images of these structures were copied on colourless, clear or frosted acetate. The image was sandwiched between clear quartz cabochons and they were passed off as dendritic or plume agate doublets.

Another kind of fraud in the world of agate collecting is misrepresentation. Experienced collectors and dealers are generally capable of recognising the source of any agate by examining external and internal features. This is something only experience can teach. Many unethical dealers and collectors may offer agates that are misrepresented as to the source. This practice is more commonly observed with agates from northern Mexico. Low grade agates from undetermined sources may be passed off as Laguna or Coyamito agates, two varieties that command high prices even for rough material. Parcels of so-called Lake Superior agate may contain high percentages of any kind of fine-grained, red rock, including jasper, basalt, rhyolite and chert.

Artificially coloured agates from Brazil are very common on the market. Aniline dyes have been used to colour agates and these have gaudy, bright orange, green, blue, purple and red tones that are only 'skin deep' for the dyes only penetrate a small distance into the fibrous structure of the agate. Some pale yellow agates are heated by covering them in sand to produce an oxygen-poor environment. This alters the chemistry of the iron compounds within, reducing the colourless iron compounds in the bands to metallic iron. Oxidised iron will produce haematite which imparts a red to reddish-brown colour to the agate. Another method for artificially colouring Brazilian agates is to soak the colourless or pale stones in a sugar solution or honey. After the sugars penetrate into the bands over several weeks, the agate is dried and dipped into concentrated sulphuric acid, which is such a water scavenger that it removes the water from the sugar molecules and leaves a carbon residue that produces a black or dark brown colour. Trublack™ agates are Brazilian agates that have been coloured artificially by a patented electrochemical or photochemical process. They first appeared on the market in 1946. Each of these agates has a small, uncoloured square on one of the surfaces where it appears that an electrode was attached. Trublack™ agates actually come in several colours such as blue, yellow or a creamy white, as well as black and white. All of the above processes create an artificial colour restricted to the surface.

Even if most agates are relatively inexpensive compared to diamonds, rubies or emeralds, there still exists the possibility of forgeries and altered stones. A collector needs to exercise extreme caution, even when purchasing low-priced stones.

Uses of agates through the ages

Artefacts

Moss agates were used for making neolithic tools such as arrow and spear points, hide scrapers, needles and awls. Few Stone Age artefacts were made from banded agates depending on local occurrence, and tools made from agatised wood were more common. Some very old agate artefacts dating back as far as 9000 years have been recovered in Mongolia, and in western Asia, in the Levant, people of the Mesolithic Natufian Culture (10,000-8000BC) produced knives or arrowheads from moss agate. Similarly, Native Americans made considerable use of moss agate and agatised wood. There are probably two reasons that account for the paucity of banded agate tools: (1) banded agates make up a small percentage of the cryptocrystalline family of minerals which make up agates, and (2) banded agates are not as easily tooled as are other forms of cryptocrystalline quartz. A possible reason for banded agate being tougher to tool is that this type of agate is made up of microfibres tightly bound into a twisted pattern making it considerably more compact than others, and therefore harder to work.

Opposite This frog was carved from bloodstone by Paul Dreher. Bloodstone is green agate containing speckles of red jasper. All these derive from the cryptocrystalline quartz group.

Above *Made from Mesopotamian seals which date from 2200 to 350BC, this jewellery set once belonged to Lady Layard, wife of one of the first excavators of ancient Assyria, Henry Layard.*

Jewellery

Agate jewellery may have had its origins in western Asia. Some of the earliest agate jewellery, from around 2500BC, has been found among artefacts from the city of Ur in Sumeria (now part of modern Iraq). Jewellery from the tomb of Queen Puabi included carnelian agate beads, whilst other Sumerian jewellery includes agate beads that have been chemically etched with soda to produce fine white lines. Sumerian agate beads have been used in necklaces as well as head bands. Agates are among the most difficult stones in which to drill holes, and some of these beads show that the early lapidaries had reached a relatively high point of technological sophistication.

Above This agate cylinder seal dates from 5th to 4th century BC Persia. Allegedly found in a tomb at Thebes in Egypt, it depicts the Persian king, Darius the Great, shooting a lion.

Right Carnelian scarab from Egypt, approximately 2.5 cm (1 in) long.

In Egypt during the Middle Kingdom (2040-1730BC) carnelian agate beads were used in jewellery items including necklaces and bracelets. The Egyptian beads of this period were often very small, and hundreds of individual beads were used in a single item. Other pieces of jewellery such as scarabs included small, cut agates that were combined with many other stones to produce fine inlay work. Pictures from Egyptian tombs show that lapidaries or gemstone cutters, bead makers and metal workers worked side-by-side in the same studios; this practice is almost unheard of in modern times. In the New Kingdom (1567-1085BC) carnelian agate beadwork was also used in necklaces and bracelets. Deep-red carnelian agate amulets, to connote the red of life's blood, were worn by ancient Egyptians to protect against misfortune.

Left An agate seal of Minoan Crete origin, showing a man leading a bull. It dates from approximately 1500–1300BC.

The Mycenaean peoples (*c.* 1400BC) who inhabited what is now modern Greece, developed elaborate gold jewellery, using carnelian agate beads as accent pieces. Grecian jewellery was often characterised by multi-coloured inlay stones, many of which were agate. One of the most important developments in Grecian agate jewellery was the cameo. Cameos probably evolved from the earlier Sumerian use of onyx agates to produce an eye or a series of concentric circles in a finished stone. Because onyx agates have multi-coloured, parallel bands, they are the most desirable stone for cameos. In early times, cameos were not usually used in jewellery but were kept as ornaments and decorations; they may have been an early example of an agate collectable.

Romans used agate in much of their jewellery. The first agates were recorded in about 350BC from the River Achates (now Dirillo) in Sicily by Theophrastus in the 4th century BC. This large Mediterranean island was under Greek domination at the time Theophrastus first described agates, and it is likely that the agate used for Grecian cameos came from this area. The Grecian provinces in Sicily were the birthplaces and homes of many of the finest engravers the world has known, and their work can be seen on both cameos and coins of that era. This workmanship on cameos generally deteriorated during Roman times but engraved stones, including cameos and intaglios of fairly high quality with Roman motifs, were taken from a late Roman tomb in Carthage (North Africa) dated at about AD400.

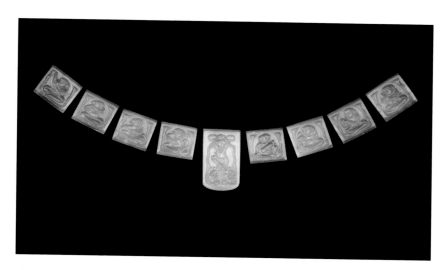

Above *An agate belt set from China, Tang Dynasty, AD618–906, featuring a set of nine agate plaques carved with foreign musicians.*

Examples of agate jewellery have been found throughout the area occupied by the Romans, e.g. a bracelet with sard onyx from Roman Egypt of the 3rd century AD, a bloodstone intaglio with Nike, the Goddess of Victory from the 2nd or 3rd century AD, and a carnelian intaglio with a scorpion from the 1st or 2nd century AD.

Agates were used in jewellery in Britain during the Roman occupation, and most of the known jewellery pieces are either cameos or intaglios. It has been suggested that many of the agates used in Roman Britain were of local origin and make up a large percentage of jewel stones of that era. Much of the rough agate material would have come from Scotland, the only really plentiful source in Britain. There are several examples of Roman era agate pieces including a hoard of 110 un-mounted, engraved carnelians that were collectively called the Snettisham jewellers' hoard. Agate cameos have also been found that show Graeco-Roman deities such as Hercules, a medusa, a lion and a bear. The engravers may have used layers of different coloured glass to imitate banded agate.

Agate was not commonly used for jewellery in China, their preference being jade, but a burial necklace for a young princess of the Sui Dynasty (*c.* AD600) contained several agate beads, and agate plaques for belt fastenings have been found with Tang Dynasty (AD618-906) artifacts.

Above *Woman's belt buckle, Kazakh, late 19th century AD from Kazakhstan.*

Eurasian nomads such as the Mongols and Turkomans used agates in much of their tribal jewellery. Many Mongol women's head-dresses carried agates that were simply attached with horse-hoof glue and a few strands of thread. Turkoman jewellery includes head-dresses, pectorals, necklaces, pendants, rings and earrings, many of which include carnelian agate. Turkoman jewellery, many pieces of which contain carnelian, has enjoyed an upswing in popularity at the time of writing and it is often seen in shops and on Internet auctions.

Right *Cup made of shaped and polished agate, with silver-gilt mounts from Germany.*

Agate 'showstones' date back as far as the Merovingian Dynasty in France (*c.* AD500) and through the Dark Ages, and have been found in graves of this era. Showstones were usually mounted in engraved silver and were prized as talismans and luck pieces. The use of showstones continued into the 17th century in England. Some agate was used in 17th century French jewellery, and contemporary carved cameos have been seen in some collections. Agate appears to have enjoyed more use in England in the 17th and 18th centuries, and some fine pieces were owned by the Duchess of Marlborough. In the late 19th and early 20th centuries, Scottish agate jewellery became very popular, and such pieces are still eagerly sought by collectors, although they are infrequently worn nowadays. The early 20th century saw the development of the Art Nouveau movement, and stained or dyed agates of single, pastel colours were popular jewellery items, in contrast to the brightly coloured Scottish agates.

In North America, agates were recorded from the original colonies that became the USA, and from maritime provinces of Canada, especially Nova Scotia. Many of these early localities have since been forgotten, and North American agate jewellery pieces from the late 18th and early 19th centuries are very rare antiques. In the USA in the l9th century, a costume jewellery industry grew up around the then abundant agate resources in the vicinity of the three northern Great Lakes: Michigan, Superior and Huron. Jewellery made from drilled, tumble polished stones became a desirable tourist item. A thriving agate-like marble industry developed in Michigan in the late 19th century and these antiques are now revered by collectors as much as precious jewellery. Western North America witnessed the development of an inexpensive costume jewellery movement in the late 19th century which still persists today. Tourist shops and jewellery stores in such areas as Montana and Oregon in the USA, and Manitoba and British Columbia in Canada have many examples of inexpensive agate pieces.

From the late 1880s through to about the onset of the First World War, rock and mineral collecting grew as a hobby. Several publications such as the *Mineral Collector* thrived, and popular accounts of gems and agates appeared in popular magazines and newspapers. A somewhat general paper was published in Britain by W.J.L. Abbott in 1887 that dealt in part with various kinds of agates, but it had no illustrations or photographs to accompany the text. Guides on agates that appeared around this time appear to have been published to be used while viewing collections in museums; an example of such a publication was J.G. Goodchild's *Guide to the collection of Scottish agates* in 1899.

Shortly before the First World War, rock and mineral collecting went into a decline, and even the appearance of Rafael Liesegang's publication *Die Achate* in 1915 did not generate much new interest in agates. This was likely a result of the ongoing war. Rock and mineral collecting enjoyed a resurgence in the late 1920s. In 1927 a popular guide to agates that dealt with their physical properties and origin by O.C. Farrington was published, and provided an interesting explanation of the archaeology and folklore of agates. This volume is still used as an important historic reference. In 1927 Peter Zodac began publication of the *Rocks and Minerals* magazine, which featured a number of special issues dedicated to agates. Several new publications appeared shortly thereafter, and the hobby of agate collecting drew many enthusiasts on a worldwide scale, a hobby which continues today.

After the Second World War, the hobbyist became an important source of agate jewellery in North America, and this movement followed a few years later in Europe when post-war reconstruction allowed time and resources to pursue such activities. In the late 1900s and early 2000s, many skilled artists have rediscovered agate as a central stone in their creations, and agate jewellery has reached a new level of acceptance by the general public. The finest pieces now turn up in art galleries rather than tourist shops.

Early writings and collections

Some people have a strong urge or instinct to collect. We do not know when people first began to collect things though some archaeological sites have yielded what appear to have been collections. Frozen tombs of Scythians (Eurasian nomads) have yielded objects of no utility on the Steppes, for example, tropical cone shells, pyrite crystals and agate intaglios. These items may all have been parts of collections. Theophrastus described agates and many kinds of rocks and minerals in about 350BC in a work he titled *Peri Lithon* or '*On Stones*'. We do not know if Theophrastus made a collection of stones, we only know that he wrote about them. To make such a lengthy manuscript, Theophrastus likely maintained at least a working collection of stones for comparative examples, but it is uncertain whether he kept them and placed them in a safe repository or if he simply dispersed them. The same can be said for Pliny the Younger who wrote about stones in the 1st century AD.

Through the Dark Ages in Europe, little writing about any subject was produced and there is little evidence for the existence of collections. During much of the Dark Ages, the Arab world was the haven for the arts and sciences, and many manuscripts about gemstones, including agates, were written by Arab scholars. As the Dark Ages came to a close, new writings on gemstones appeared in both Arabic and European works. In North Africa and nearby areas in what is now Israel, Syria and Iraq, Ahmed ibn Yusuf al Tifaschi produced the volume, *The Best of Thoughts on the Best of Stones* and he described several kinds of agates. Alfonso X el Sabio (Alfonso the Wise), King of Spain, AD1226-1284 wrote a 'lapidary', a treatise on stones, describing many stones, including agates. At about the same time in Germany, Albertus Magnus wrote a lapidary and described several kinds of agates. There appears to have been a lack of information exchange relating to agate sources and scientific findings between Europe and the Arab countries during this time, similarly during the Cold War years (1945-1990) between the Soviet Union and Europe.

The first documented collection that included many agates may have been made by Lorenzo de Medici (Lorenzo il Magnifico) in Florence, Italy, in about 1450. Much of Medici's collection is still intact in the museums in Florence and Naples. Medici collected fine, finished gems, statuary made from precious stones, cameos, as well as examples of items that are common, such as polished slices of agate and flint. Medici's collection showed not only an appreciation for beautiful objects of art, but also an appreciation for all rocks and minerals. During the Napoleonic Wars (1803-1815) the popular German sources of agate were cut off, and people turned to local sources of agate. In Scotland, mass production of jewellery was initiated by a local Scotsman Gavin Young, and local collectors would sell their finds to the jewellery houses.

In the early 1970s, natural objects such as fossils, crystals, fossilised wood logs and agates became popular household decoration items in Europe and the USA. Many of these kinds of items also found their way into pop culture and movie and TV set decorations. Many fine agates can be seen decorating the space ship, *Enterprise*, in the series *Star Trek™: The Next Generation*. This probably arose from the fact that the series creator, Gene Roddenberry, was a skilful and ardent lapidary.

The use of agates in science and technology

Agates are not generally thought of as having a great deal of scientific or technological application, but there have been a few areas where agate has been used. The agate mortar and pestle was used in chemistry and physics laboratories for many years. The smooth, highly polished surfaces of the agate allowed the scientist or technician to grind chemical reagents. The tools would remain clean and they enabled the chemist to work to very high standards of accuracy. In addition, agate is resistant to acid, so cleaning such a tool to chemical purity is perfectly possible.

Highly polished agates have little friction, so bearings for analytical balances of the chemistry and physics laboratories of the past were made from agate. Such balances allowed accurate weighing to increments as small as 1/10,000th of a gram.

Below Pestle and mortar made of agate.

Turn and bank indicators of military aircraft used to contain a black agate bead. Some 400 tonnes of agate from Queensland, Australia, were mined during the Second World War to provide stones for this use and other military and scientific uses. The position of the bead in a liquid medium enclosed in a glass tube told the pilot when his aircraft was level; such technological applications enabled pilots to make safe landings at night on both land and on aircraft carriers.

The application of agates in exploring modern geological problems is now in its infancy. In Finland, geologists K.A. Kinnunen and K. Lindqvist were shown some agates that had peculiar 'chicken-wire' fracture patterns. The origin of these agates was questioned as they had been found rather unusually in glacially derived sediments. Upon tracing these agates back to their original source, Kinnunen and Lindqvist found that the strangely fractured agates were associated with ancient meteor impact sites. They have christened such fractured agates as 'pathfinders' for the important role that they played in the discovery of evidence for ancient impact areas.

In England, T. Moxon has conducted research into crystal sizes in wall-banded agates. Moxon plotted the crystal sizes of these agates against the logarithms of the geological ages of their host rocks. He observed a general increase in the crystal sizes with the increasing age of the host rock. Moxon found a high correlation between the agate crystal sizes and the ages of the host rock, and he suggested from these observations that the agates formed almost contemporaneously with their host rock. Moxon went on to suggest that crystal sizes of wall-banded agates may provide approximate ages for their host rocks.

In Scotland, A.E. Fallick and others suggested that water bound to microcrystals making up the agate have retained their hydrogen isotope ratios since the time of their deposition. They compared oxygen isotope ($^{18}O/^{16}O$) and hydrogen isotope (D/H)1 ratios of the bound water in thirteen agates from the lower Devonian of Scotland (410 million years old) and the seven agates from Tertiary of Scotland (62 million years old). The isotope ratios differed in two samples, but they were able to correlate them linearly with isotope ratios in modern meteoric (rain) water. Their data showed that water content in agate has some genetic significance, and they further suggested that agates formed at temperatures of about 50°C from fluids containing some meteoric waters.

In Germany, G. Holzhey has actively pursued studies on thunder egg agates, a kind of agate that forms in rhyolitic rocks (see p. 49) from the Black Forest region in Thuringia. Holzhey's work has shown differences in chemical makeup of the outer and inner portions of the shell that surrounds the core of the thunder egg. M. Landmesser, also in Germany, has done important theoretical work on the mobility and transportation of silica that makes up banded agate.

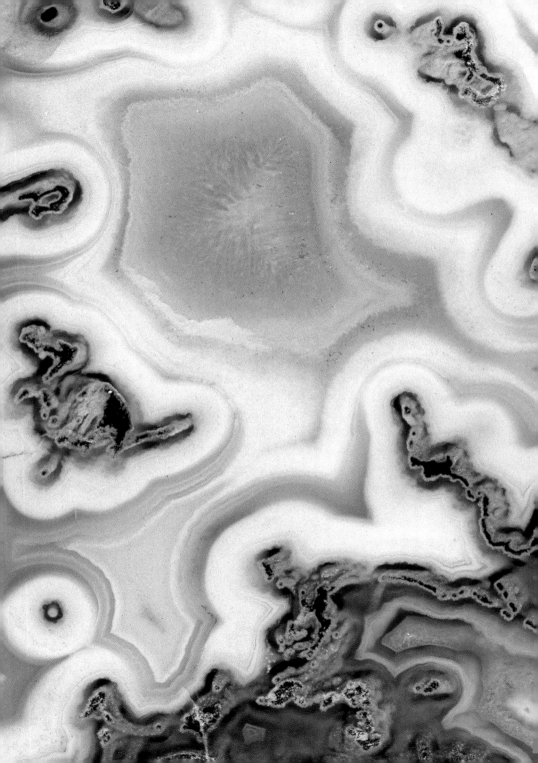

Collecting

It is likely that the first agate any collector had in his or her collection was not found by that person. Most likely, they received the specimen from another collector, or that they purchased it at a rock shop, rock show or tourist attraction. It is even likely that the soon to be collector had no idea that one might be able to go out and search the open fields, rock outcrops or gravel deposits in streams for agates.

Preparation

Before beginning a search for agates in the field, there are a few things you need to do and know. In most instances, you will be driving to the agate collecting areas. You need to be sure that your vehicle is field-worthy and capable of negotiating primitive roads or trails. Even in well-developed areas, some very primitive roads exist, and these are the ones that lead to the steep terrain where you are most likely to locate agates.

Do get maps that define the collecting area very well. In Britain, Ordnance Survey maps should provide the necessary details, whereas in the USA, US Geological Survey topographic maps will usually provide the necessary details of roads, streams, outcrops, pits and quarries.

Keep in mind that even if you see a potential collecting area, it may not be permissible to collect there. The land may be privately owned which is almost certain to be the case in Britain, in which case permission is needed. Public lands may be situated in parks or reserves of some kind, where collecting of anything is not allowed. In Britain collecting is not allowed in any of the national parks. In the USA, some mines or quarries have found agates or other mineral specimens to be a profitable sideline, and have leased their sites to commercial collectors. Some localities, especially in the USA and Canada, are fee localities where collectors can pay a daily or a poundage fee for what they collect.

Opposite A cut and polished section of blue agate from Idar-Oberstein, Germany.

In both Britain and the USA, responsible groups of collectors as well as scientists have developed codes of ethics for collecting rock, mineral and fossil specimens in the field. In England this is called the Countryside Code, in Scotland the Scottish Outdoor Access Code and in the USA the American Federation of Mineralogical Societies Code of Ethics. Such codes exist in almost every country and they should be followed strictly. Keep in mind that such codes are developed to meet the needs of each specific country. What may work in Britain may not work in the USA. If you travel to another country to collect, be sure to obtain the code of ethics as well as the legal codes that govern collecting there.

In the field

The kind of site you visit will determine the tools that you will need for a successful collecting trip. If you are going to collect along a gravel bar in a stream or an outcrop of weathered glacial till, you will usually be able to do all of your collecting with a small wedge (or chisel), a spray bottle of water and a canvas bag to carry your specimens. Most of the agates will be loose, and in unconsolidated debris. These agates have been eroded from their source and transported to where they are found by streams or glaciers. The real trick is to be able to recognise an agate when you see it. Many fine specimens remain in the field because even advanced collectors do not readily recognise them, especially if there are no apparent bands. Many agate nodules have a waxy or greasy surface and many are somewhat translucent, and the surfaces of the nodule are often pockmarked. Deeply weathered agate nodules, however, are often chalky white on the outside, but the pockmarks remain.

When working gravel bars or glacial till, you should always look over the outcrop from several directions. Light and shadow vary considerably, even on relatively clear days, and a line that you walked a few moments earlier and produced nothing may produce an agate when you are looking along it from the opposite direction.

If you are searching for agates in their host rock at a mine or quarry, you will need to use a completely different set of tools than for outcrops of unconsolidated sediments. You will need several sizes of hammers, pry-bars, gads and chisels. You should also have safety goggles and a hardhat when working in mines or quarries. Be sure that your chisels are properly sharpened and the striking end is free of burrs – these can be removed with a grinder or file. These burrs can fly off the end of a chisel with the velocity of a rifle bullet and cause very serious injury.

Some countries require that anyone working in mines or quarries undergo a training session on mine procedures and safety, and require that all workers in the mines or quarries wear protective gear. Some mining or quarrying companies require proof of financial liability before allowing a collector on their property.

Regardless of the kind of outcrop you visit: a ploughed field, or a quarry, you should always dress appropriately for the conditions. Wear heavy shoes or boots, long trousers and long-sleeved shirts to protect yourself from the wind, rain and sun.

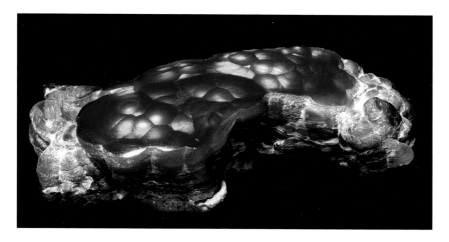

Above *Fire agate from Aguascalientes, Mexico.*

Further information

Journal references cited

Cameron, W.E., 1900, The Etheridge and Gilbert gold fields, *Geological survey of Queensland*, report no. 151.

Coquand, G., 1848, Description des terrains primaires et ignés du departement du Var: *Mémoires de la Société Géologique de France* (2), v. 3.

Czaja, M., 1992, Results of structural studies of quartz and chalcedony from the lubiechowa basaltoids (The Kaczawa Mountains), *Prace Mineralogiczne* 82, Polska Akademia Nauk, Oddzial W Krakowie, Komisja Nauk Mineralogicznych.

Frazier, S. and Frazier, A., 1988a, Name that agate, *Lapidary Journal*, v. 42, no. 1, p. 65-77.

Freyberg, B.V., 1927, Observaciones geológicas en la región de las ágatas de la Serra Geral (Rio Grande do Sul, Brazil), *Boletín de la Academia Nacional de Ciencias de la República Argentina*, v. 30, p. 129-170.

Heaney, P. J. and Davis, A. M., 1994, Geochemical self-organization in agates at the sub-micro scale, *Geological Society of America, Abstracts with Programs*, v. 26, p. A-111.

Kay, J.R., 1980, The geological story of the Cedar Creek thunder egg deposit at Thunderbird Park Tamborine Mountain: Queensland 4272, *Geological survey of Queensland*, Rec. 32, p. 1-15.

Kinnunen, K.A. and Lindqvist, K., 1998, Agate as an indicator of impact structures: An example from Saaksjarvi, Finland. *Meteorites and Planetary Science*, v. 33, p. 7-12.

Landmesser, M., 1984, Das Problem der Achatgenese: Mitteilungen Pollichia, v. 723, p. 5-137.

Lane, A.C., 1911, The Keweenaw Series of Michigan, *Michigan geological and biological survey publication* 6, 983 p.

Leiper, H., The agates of North America, *The lapidary journal*, 1966.

Mattos, L. E. de, 1974, Perfil analitico da agata, Ministéro das Minas e Energia, *Brazil Departmento Nacional da Produção Mineral, Boletim* 29, 18 p.

Raisin, C., 1889, On some nodular felstones of the Lleyn, *Quarterly journal of the geological society of London*, v. 44, p. 247-268.

Rampino, M.R. and Stothers, R.B., Flood basalt volcanism during the past 250 million years. *Science*, 241, 1988.

Strom, R. N., Upchurch, S. B. and Rozenweig, A., 1981, Paragenesis of "box-work geodes," Tampa Bay, Florida, *Sedimentary Geology*, v. 30, no. 4, p. 275-289.

Books

Agate collecting in Britain, P.R. Rodgers. Batsford, London and Sydney, 1975.

Agate, Microstructure and Possible Origins, T. Moxon. Terra Publications, 1996.

Agates, H. Macpherson. British Museum, (Natural History), London, and National Museums of Scotland, 1989.

A History of Jewellery 1100-1870, J. Evans. Boston Book and Art Publisher, Boston, MA, 1953.

Ancient Greek and Roman Gold Jewellery in The Brooklyn Museum, P.F. Davidson, and A. Oliver. Brooklyn Museum, Brooklyn New York, 1984.

Arab Roots of Gemology, S.N. Abul Huda. Scarecrow Press, Lanham MD and London, 1998.

Banded Agates: Origins and Inclusions, R.K. Pabian and A. Zarins. Conservation and Survey Division, IANR, University of Nebraska-Lincoln, 1993.

Catalogue to the Jewellery, Greek, Etruscan, and Roman, in the Departments of Antiquities, British Museum, F.H. Marshall.University Press, Oxford, 1969.

Desert Gem Trails. A Field Guide to the Gems and Minerals of the Mojave and Colorado Deserts, M.F. Strong. Gem Guides Book Co., Baldwin Park, CA, 1971.

Dictionary of gems and gemology, R.M. Shipley. Gemological Institute of America, Santa Monica, California, 1971.

Die Achate, R.M. Liesegang. Verlag von Theodor Steinkopf, Dresden and Leipzig, 1915.

Ellensburg Blue, J.P. Thomson. Spokane, Washington, 1961.

Gem and Lapidary Materials for Cutters, Collectors, and Jewelers, J.C. Zeitner. Mountain Press Publishing, Misoulla, Montana, 1996.

Flint, its origin, properties, and uses, W. Shepherd. Faber and Faber, London, 1972.

Gem Trails of Arizona, B. Simpson and J.R. Mitchell. Gem Guides Book Co., Baldwin Park, CA, 1989.

Gem Trails of California, J.R. Mitchell. Gem Guides Book Co., Pico Rivera, California, 1986.

Gems: their sources, description, and identification, R. Webster. Butterworth's, London, 1970.

Gemstones of North America, J. Sinkankas. Van Nostrand Reinhold Company, New York, Toronto, London, Melbourne, 1959.

Greek and Roman Jewellery, R.A. Higgins. University of California Press, Berkeley and Los Angeles, 1980.

Guide to the collection of Scottish Agates, J.G. Goodchild. Royal Scottish Museum, Edinburgh, 1899.

Jewellery: An Historical Survey of British Styles and Jewels, N. Armstrong. Butterworth Press, Guilford and London, 1973.

Jewellery 7000 Years, An International History and Illustrated Survey from the collections of the British Museum, H. Tait. Harry N. Abrams, New York, 1986.

Minéralogie, ou, description générale des substances du règne mineral, Johan Gottschalk Wallerius. Paris, 1747.

Mongolian Geology, Z. Baras, M. Togooch, T. Batjargal, D. Sengee and K. Baasanjav. State Publishing House, Ulaanbaatar, Mongolia, 1989.

Mongol Jewellery: Carlsberg Foundation's Nomad Research Project, M. Boyer. Thames & Hudson, New York, 1995.

Old silver jewellery of the Turkomah: an essay on symbols in the culture of Inner Asian Nomads, D. Schletzer and R. Schletzer. Reimer, Berlin, 1984.

On stones; introduction, Theophrastus. Ohio State University, 1956.

Precious Stones, M. Bauer. Charles E. Tuttle Co., Publishers, Rutland, Vermont and Tokyo, Japan, 1896.

Rockhounding Nevada, W.A. Kappele. Falcon Publishing Company, Helena, Montana, 1998.

South Dakota's Fairburn Agate, R. Clark. Palmer Publications, Amherst, 1998.

Structural Model of Italy (sheet 6), map and text, G. Bigi and others. Consiglio Nazionale delle Recerche, Projetto Finalizzato Geodinamica, Selca, Florence, Italy, 1990.

Terminal Cretaceous Environmental Events, C.B. Officer and C. L. Drake. Science, 227, 1985.

The Agate Book, Including a description of Agate Filled Thunder Eggs, a Handbook for the Agate Collector and Cutter, H.C. Dake. Mineralogist Publishing Company, Portland, Oregon, 1951.

The Agates of Northern Mexico, B. Cross. Burgess Publishing, Edina MN, 1996.

The Beauty of Banded Agates, an Exploration of Agates from Eight Major Worldwide Sites, M. Carlson. Fortification Press, Edina MN, 2002.

The Formation of Agates, W. J. L. Abbott. *Proceedings of the Geological Association,* London, 1889.

The Formation of Thunder eggs, R. Colburn. Basin Range Volcanics Geolapidary Museum, Deming, New Mexico, 1997.

The Jewellery of Roman Britain, C. Johns. University of Michigan Press, Ann Arbor, Michigan, 1996.

The Lake Superior Agate, S. Wolter. Burgess Publishing Company, Edina, Minnesota, 1994.

The Mineralogy of Scotland, Volume I, M.F. Heddle. David Douglas, Edinburgh, 1901.

The Pebbles on the Beach, C. Ellis. Faber and Faber Limited, London, 1954.

The World of Jewel Stones, M. Weinstein. Sheridan House, New York, 1958.

Victorian Jewellery, G.R. Dawes and C.I. Davidov. Abbeville Press, New York, 1991.

Western Asiatic Jewellery 3000-612BC, K.R. Maxwell-Hyslop. Methune & Co. Ltd., London, 1971.

Journals

Australian Gemmologist, Gemmological Society of Australia, Sydney.

Lapidiary Journal, Malvern, PA, USA.

Lapis, Christian Weise Verlag, Munich, Germany

Rock and Gem, Miller Magazines Inc., Ventura, CA, USA

UK Journal of Mines and Minerals, Donaster, UK.

Websites

NB. Please note that website addresses are subject to change.

Canada: http://www.canadianrockhound.ca/about.html

Iran: http://iranian-agates.freeservers.com/

Netherlands: http://agates.freeservers.com/

Nova Scotia: http://www.rockhounds.com/rockshop/isfeld2.html

Sardinia: http://www.angelfire.com/linux/agatesdellasardegna/

Scotland: http://www.curriehj.freeserve.co.uk/agates.htm

General: http://www.minerant.org/collectors.html

http://csd.unl.edu/

Glossary

alkaline lake a lake having alkaline water, with a pH (concentration of hydroxyl or OH- ions) greater than 7.

amygdaloid commonly almond-shaped to sub-spherical nodules that form in basaltic or andesitic rocks.

andesite a dark coloured, silica-deficient volcanic rock containing abundant calcic feldspar, with biotite mica and amphibole mineral

artefact a simple form of primitive tool or art that is produced by hand.

breccia a kind of rock that is composed of fragments of a pre-existing rock derived from movement along a fault or volcanic activity.

cabochon a non-faceted gemstone; that is, one with a curved cross-section and no flat surfaces.

crossed polarisers a situation where the moveable analyser of a polariscope is at right angles to the stationary polariser; i.e. the field will appear dark. In uncrossed polarisers, the field will appear light.

crystal a polyhedroid with smooth, planar surfaces that reflect the internal arrangements of atoms making up the crystal.

dendrite tiny tree- or bush-like inclusions that are usually composed of manganese or iron oxides.

doublet usually, a cabochon cut gemstone that has a very thin layer of material that is cemented to a cap of clear quartz or glass, and is frequently used for plume agates that may be nearly opaque in thick slices.

fortification agate a banded agate, so called because the bands resemble what is seen in an aerial view of ancient or medieval fortifications. The term is not commonly used nowadays.

fossil direct (shells, bones, wood) or indirect (tracks, burrows) evidence of ancient life forms.

lapidary (old) a book or treatise dealing with gemstones or even common stones.

lapidary (new) a person who shapes and finishes gemstones and other stone objects.

nodule a stone that has formed in a bounded cavity such as a gas bubble in volcanic rocks or an animal burrow in sedimentary rocks.

paint a non-scientific term of questionable value that is used to describe opaque, pastel coloured agates that have been deeply weathered.

polarised light a beam of light in which the waves vibrate in a single direction only.

polymorphs different structural modifications of the same compound e.g. crystalline quartz, moganite and common opal are polymorphs of silicon dioxide.

rhyolite a light coloured, finely textured, silica enriched volcanic rock that shows evidence of plastic or liquid flow

showstone an agate in an engraved silver mounting that was carried as a talisman or good luck piece.

silica silicon dioxide, SiO_2; more commonly applied to silicon dioxide that is in transport as a colloid or in solution than to crystalline quartz or agate.

spherulite tightly packed groups of crystals that radiate from a central nucleus.

sunbursts manganese or iron oxide inclusions that superficially resemble the solar corona or sunspots.

tholeitic a usually continental flood basalt extruded from long fissures in the earth and containing labradorite feldspar with pyroxene minerals and small amounts of glass

thunder egg a kind of agate nodule that is characterised by having a square to star-shaped central cavity that is filled with opal or agate and has a matrix of rhyolitic ash.

tuffs a porous usually stratified (layered) high-silica volcanic rock consisting of glass and ash particles that are fused together.

Uruguay banding a term used by Landmesser (1984) for parallel, straight or flat bands in an agate. Also called onyx banding or level banding.

Volcanic breccia a rock derived from explosive eruptions that is made up of fragments greater than 2 mm in diameter.

zeolite a large family of hydrous calcium, aluminium or sodium silicate minerals that have similar compositions and have usually formed as alteration products of igneous rocks.

Index

Acknowledgements

Without the aid of several other individuals, the completion of this publication would not have been possible. Trudy Brannan has provided a great deal of commentary, and Hillary Smith has sourced many of the images herein. Thanks to Harry Taylor for photographing many of the specimens. Thomas White has supported the photographic efforts in the USA as well as made available many specimens. Beth M. Wilkins enabled photography of archeological specimens at the University of Nebraska State Museum, Lincoln. Russell and Doris Kemp and Dorothy Ascher of the Lizzardo Museum of Lapidary Art, Elmhurst, Illinois, provided images of cameo replicas and Scottish Victorian jewellery. Fred B. Holbert, Robert M.B. de Jager; Brian Isfeld, Duane Mohlman, Maziar Nazari, Linda Plock, Hideharu Yamada, Jean-Baptiste Silla and Stephanie Gregoire have generously shared knowledge, images and literature that have filled many gaps in the material at hand. Brian Jackson, Peter Tandy and John Cromartie have suggested ways and words to enhance both the readability and accuracy of this book.

Picture credits

pp. 160, 161 top, 162-64 © The Trustees of the British Museum.

pp. 4, 12, 13, 28, 29 right, 34, 55, 62-71, 144, 151-55 © The Trustees of the National Museums of Scotland.

pp. 24, 25 bottom, 26, 27, 39 left, 30, 31 bottom, 35, 39, 40, 42, 43, 49, 50, 51 top, 56, 76, 78-81, 83-6, 88-94, 96-108, 110-11, 113-120, 124-34, 136, 139, 140, 142, 146, 147 © Natural History Museum. Specimens owned by Roger Pabian.

p. 161 bottom © Natural History Museum. Specimen from the Anthropology Division of the University of Nebraska State Museum, UNSM A95.07.235.

pp. 52 bottom, 59, 77, 82, 87, 95, 109, 112, 123, 145 © Natural History Museum, drawn by Lisa Wilson. pp. 14-15 © Natural History Museum, drawn by Lisa Wilson, redrawn from Agates by H.G. Macpherson, 1989, pp. 15-17 © The Trustees of the National Museums of Scotland.

p. 25 top © Ray Simons/Science Photo Library. pp. 33 bottom, 36, 47, 58 © Dirk Wiersma/Science Photo Library.

pp. 138, 143, 165 © V&A Images/ Victoria & Albert Museum.

All other images are copyright of the Natural History Museum and taken by the Image Resources Photographic Unit. For copies of these and other images, contact the Picture Library directly at the Natural History Museum or view their website at www.nhm.ac.uk/piclib.

Every effort has been made to contact and accurately credit all copyright holders. If we have been unsuccessful, we apologise and welcome corrections for future editions or reprints